Nontraditional Security Challenges in Southeast Asia

Nontraditional Security Challenges in Southeast Asia

The Transnational Dimension

Amy L. Freedman
Ann Marie Murphy

LYNNE RIENNER PUBLISHERS

BOULDER
LONDON

Published in the United States of America in 2018 by
Lynne Rienner Publishers, Inc.
1800 30th Street, Boulder, Colorado 80301
www.rienner.com

and in the United Kingdom by
Lynne Rienner Publishers, Inc.
3 Henrietta Street, Covent Garden, London WC2E 8LU

Library of Congress Cataloging-in-Publication Data
Names: Freedman, Amy L., author. | Murphy, Ann Marie, author.
Title: Nontraditional security challenges in Southeast Asia : the transnational dimension / by Amy L. Freedman and Ann Marie Murphy.
Description: Boulder, Colorado : Lynne Rienner Publishers, Inc., 2018. | Includes bibliographical references and index.
Identifiers: LCCN 2017037269 | ISBN 9781626376854 (hc ; alk. paper)
Subjects: LCSH: Security, International—Southeast Asia—International cooperation. | National security—Southeast Asia—International cooperation. | Emergency management—Southeast Asia—International cooperation.
Classification: LCC JZ6009.S644 F74 2018 | DDC 355/.033059—dc23
LC record available at https://lccn.loc.gov/2017037269

British Cataloguing in Publication Data
A Cataloguing in Publication record for this book
is available from the British Library.

Printed and bound in the United States of America

The paper used in this publication meets the requirements of the American National Standard for Permanence of Paper for Printed Library Materials Z39.48-1992.

5 4 3 2 1

Contents

Acknowledgments

Our interest in transnational security issues developed as we watched a wide range of collective problems become increasingly important in Southeast Asia. As the challenges of climate change, emerging infectious diseases, migration, and food and water security rose on the domestic and foreign policies agendas of Southeast Asian countries, we began exploring what were for us, traditionally trained political scientists, new topics. This project evolved over a number of years as we attended conferences, presented papers, and organized panels on transnational security issues. We realized that there was a great deal of overlap in the research we were tackling separately. So, why not join forces? We want to thank the many people who were involved in these panels and shared their expertise. Their helpful ideas, feedback, and intellectual give and take helped keep us engaged and improved our work. There are many people whose support, comments, and critiques of early versions of our manuscript must be gratefully acknowledged. This is probably an incomplete list, but the following merit thanks by name: Lee Jones, Nick Thomas, Tom Pepinsky, Liz Wishnik, Sophal Ear, David Denoon, Don Weatherbee, Bridget Welsh, and Kai Otswald. Likewise, we want to thank the external reviewers, as well as Carrie Broadwell-Tkach at Lynne Rienner Publishers for her confidence in the project, her patience, and her support.

In addition, Amy would like to thank her Long Island University colleagues for their support and Anna Gallo, department secretary, and Peter Grabowski, student assistant, for their help with some of the final steps in the manuscript preparation. At Seton Hall, Ann Marie would like to acknowledge the support of a University Research Council grant and conference funding from the School of Diplomacy and International Relations. She would also like to thank her graduate students Alyssa Pack, Sarah Ireland, Morgan McMichen, and Meagan Torello for their expert research assistance over the years and Susan Malcolm, the school secretary, for her invaluable administrative support.

A coauthored book is a triumph of cooperation, persistence, and patience, and ultimately a test of friendship. We are grateful that our professional and personal friendship has survived this project, and indeed it made for a stronger book. As two working moms we consulted with each other across continents and time zones; we talked on the phone while making dinner, driving kids, commuting, walking the dog, and of course at our desks. Amy offers a heartfelt thank you to her kids, Eric, Alyssa, and Cameron, and husband, Kevin, who put up with evenings and weekends of diverted attention. Similarly, Ann Marie is grateful to her husband, Jim, for juggling the demands of a busy household and his own professional obligations during overseas research trips, and to her children, James and Caleigh, for putting up with long absences and welcoming her back with open arms. We dedicate this book to them.

1

Nontraditional Security Challenges in Southeast Asia: The Transnational Dimension

While humanity has succeeded in making great technological advances and life expectancy has skyrocketed over the last half-century, we may again be faced with Malthusian-type problems of food and water insecurity, calamitous consequences from climate change, the rapid spread of new and deadly diseases, and increased migration. There is a clear link between these problems. Climate change is increasing environmental stress by producing more extreme weather and creating wetter conditions in some places and drought conditions in others. This in turn decreases the amount of arable land for agriculture, reducing food availability, raising prices, and negatively impacting food security. Climate change also threatens water security, spurs the spread of conditions leading to increased incidence of disease, and causes new migration patterns. Furthermore, all of these challenges cross state borders, demonstrating that what one country does at home impacts other countries abroad. The transnational nature of these threats means that any effective solution to them requires international cooperation.

Southeast Asia currently faces escalating challenges relating to climate change, food security, water security, migration, and the spread of infectious diseases. This book focuses on these five problems. We ask: What efforts are under way in the region to address these shared concerns? Are regional or international organizations able to facilitate cooperation on these common problems? And why or why not? We find that although states recognize that cooperation would produce better policy outcomes, they often find it difficult to do so because of conflicting interests, concerns over sovereignty, and weaknesses in governance systems within and between countries.

Overview of Transnational Threats in Southeast Asia

Climate change produced by global warming is a worldwide phenomenon that has diverse impacts on countries across the world. According to the US Environmental Protection Agency (EPA), *climate change* refers to any significant change in the measures of climate lasting for an extended period of time such as major increases in temperature, precipitation, and wind patterns.[1] In recent years climate change has primarily been produced by *global warming,* which refers to the ongoing rise in global temperatures caused by increasing concentrations of greenhouse gases (GHGs) in the atmosphere as a result of human activity. The earth's average temperature has risen by 1.5 degrees F over the past century and is projected to rise another 0.5 to 8.6 degrees F over the next hundred years.[2] Global warming has produced changes in rainfall, droughts, more severe heat waves, melting ice caps, and changes in oceans across the globe.

All Southeast Asian countries are vulnerable to the effects of climate change and, therefore, should have a shared interest in combating it collectively. Rising sea levels threaten to submerge cities such as Bangkok and Jakarta, which are prone to four climate-related hazards: droughts, floods, landslides, and sea-level rise.[3] The intrusion of saltwater into Vietnam's agricultural land is reducing productivity and negatively impacting food security. Changes in rainfall have produced droughts, such as the horrific one in the Mekong region in 2015–2016, that threaten water and food security. Rising temperatures have produced new threats from invasive insects such as the caterpillar explosion in Java and Madura in 2011 as well as loss of cultivable land that can lead to further food insecurity.[4] A key driver of global warming in Southeast Asia is deforestation. In addition to contributing to climate change, deforestation is contributing to loss of biodiversity, which poses problems for indigenous communities who rely on these ecosystems for their livelihoods, and it is producing transboundary haze caused by forest fires on peatland. Given the magnitude of the threat that climate change poses to all Southeast Asian countries, they have a strong interest in working collectively to mitigate it; yet they do not always do so.

The World Health Organization (WHO) defines "food security" as existing "when all people at all times have access to sufficient, safe, nutritious food to maintain a healthy and active life." Food security is built on the three pillars of food availability, access, and use. *Food availability* means that sufficient quantities of food are consistently available. *Food access* refers to affordability, or having sufficient resources to obtain appropriate food for a nutritious diet. *Food use* means having the appropriate knowledge of basic nutrition and care, adequate water, and sanitation to ensure a healthy diet.[5] Actions taken by governments to ensure food security for their own citizens, such as bans on exporting food, purchases of food for reserves, or incentives for farmers to plant commercial rather than food crops, can threaten the food security of people elsewhere. Finding collaborative solutions to address the basic human need of food security is a critical task.

Many factors and trends are emerging that threaten Asian and global food security. The number of people who are *food insecure* in Asia has been on the rise. While strong economic growth has bumped a number of states toward middle-income status, a large number of the world's undernourished live in Asia, where India accounts for 43 percent of the undernourished and China for 24 percent. The region's population is projected to increase from 3.6 billion to 4.5 billion people. Most of that growth will occur in urban areas, as the urban population is set to surpass the rural population as early as 2028.[6] At the same time, there has been a global decline in investment in agriculture, from a 20 percent share of official development assistance for agriculture in 1979, to 5 percent in 2007.[7] There is also a paradox: economic growth and prosperity are unquestioningly good things; however, as societies become wealthier, they consume more animal products. Thus, there is increased pressure on agriculture to produce food for livestock instead of for human consumption. This can result in increased prices for grains, which further hurts poorer citizens and areas and further strains food supplies.

Southeast Asia is a diverse region that continues to enjoy sustained economic growth and poverty reduction. Yet over 15 percent of Asia's undernourished population lives in Southeast Asia. The region has abundant natural resources and is rich agriculturally, but environmental stresses and changing demographics, lifestyles, and eating habits threaten many key ecosystems and thus pose a threat to future food production. Historically, food shortages have been a result of poor social, economic, and political policies, often combined with environmental problems of drought, floods, and blight that can exacerbate unequal allocation of resources. Today, the most probable cause of food insecurity for millions of people around the world is changing demographics coupled with the effects of climate change. The ability of societies and governments to effectively cooperate to deal with the effects of climate change on food security poses a significant challenge in Southeast Asia.

Water is fundamental to human life, and a country's water security is defined by the following benchmarks. First, household water and sanitation needs are met in all communities. Second, water supplies are sufficient to support productive economies in agriculture, industry, and energy. Third, there is ample water to develop vibrant livable cities and towns. Fourth, healthy rivers and ecosystems are restored. Fifth, communities are resilient and able to adapt to change.[8] Water insecurity, like food insecurity, results from a combination of environmental factors coupled with poor government planning or misalignment of priorities with longer-term needs.

Despite the perception that Southeast Asia is lush, rainy, and dotted with rain forests, it is water insecure. There are a number of water-related challenges in Southeast Asia: urban flooding, the need for greater water infrastructures to provide citizens with clean drinking water and water treatment capacity, the depletion of groundwater and aquifers, and the shared used of common waterways such as oceans and rivers. Actions taken by the governments, citizens, or businesses in a country can threaten the water security of others. Governments may divert waterways to irrigate their own agricultural areas, thereby reducing the water flow to other countries. Similarly, the decision by upstream countries to build dams on rivers can negatively impact that water flow to downstream countries. Water pollution by private enterprises into lakes, oceans, and rivers can poison the water on which others depend. Finding ways to cooperate and ensure an equitable distribution of such vital resources is a critical challenge in Southeast Asia.

Emerging infectious diseases (EIDs) pose international security threats the world over. EIDs are illnesses that have the potential to inflict harm on humans, crops, and livestock. And they can do significant damage to countries' health infrastructures and overall economies, as the Zika virus demonstrated in Latin America and the Caribbean in 2016 and as the Ebola virus outbreak in West Africa had clearly done to Sierra Leone and Liberia in 2014. Other examples of this were seen with the severe acute respiratory syndrome (SARS) outbreak in Hong Kong and China in 2003, and the Influenza A virus A/HIN1, otherwise known as swine flu, impact on the Mexican economy in 2009 when it was estimated that almost 1 percent of that country's gross domestic product (GDP) was lost due to fears of the virus.[9]

Developing countries, like most of those in Southeast Asia, are more likely to suffer from rapid and deadly disease transmission because of weak public health infrastructures, outdated technology, and a lack of transparency and government accountability. Countries in the region vary greatly in their EID surveillance capabilities and treatment protocols. A highly pathogenic strain of avian influenza (H5N1) poses one of the greatest risks to human health in the region, and how well or how poorly countries have addressed this threat demonstrates the challenges of confronting EIDs in the developing world. H5N1 is spread through both wild bird flocks and farm-raised flocks, and it has a high mortality rate in people infected by it. Some countries in Southeast Asia, such as Malaysia and Singapore, have been successful at detection and prevention of human outbreaks. In contrast, other countries, such as Indonesia and Cambodia, have struggled to keep outbreaks to a minimum. Neither wild birds nor deadly pathogens respect national borders. Therefore, one country's poor response to a disease outbreak not only can threaten its own citizens, but also people in other countries. Thus, it is imperative that there be national and regional efforts to combat H5N1 in Southeast Asia.

The movement of people across borders is a significant transitional issue that holds the potential for both conflict and cooperation. According to the International Organization for Migration (IOM), "migration" is the movement of people across borders or within a state and can encompass refugees, economic migrants, displaced peoples, and people moving for other purposes including family reunification.[10] Southeast Asia has long been a region with a high degree of migration. Famines in China traditionally sent informal exoduses of Chinese to Southeast Asia, and in the colonial era the British imported large numbers of Chinese and Indian workers to their colonies in Burma, Malaysia, and Singapore. Since independence, transnational migration has occurred when populations have been forced from their homes because of state and nonstate violence, or when economic opportunities are vastly better elsewhere. Southeast Asian governments have been sensitive to the security ramifications of refugees and illegal migration because of the potential for racial and religious tensions in the region's multiethnic states. Migration therefore poses a threat to the efforts of political elites to forge social cohesion and political stability.[11]

In theory, labor migration should be an area for cooperation between states. States like Malaysia and Singapore where labor is scarce can benefit economically by permitting economic migrants from labor-rich countries like Indonesia and the Philippines to work in their countries. Indeed, the mutual benefits at the heart of this exchange form the basis for many bilateral formalized efforts to regulate migration. Capturing these benefits, however, is problematic, particularly when workers and their employers dispute an issue and governments are asked to protect both sides. Similarly, Southeast Asia is home to the Rohingya minority in Myanmar, which the United Nations (UN) has called the most persecuted people on earth. As Rohingyas flee their homeland in search of safety, they often enter other countries illegally. Finding ways to manage the flow of people across borders in a manner that treats migrants humanely without negatively impacting the economic and security interests of other states and their citizens is a key challenge.

Defining Core Concepts

The title of this book refers to transnational threats. We use the term *transnational* to convey the core essence of these types of problems. Climate change, food and water security, migration, and the spread of disease are all issues that transcend national boundaries. While the globe may be carved up into territories on a map that we call states and while these states may have unique properties and interests, the problems that we address in this book are ones that do not respect state borders. They are issues whose management and resolution require states to act independently at home and collectively abroad. Terms such as nontraditional security, transnational threats, transboundary issues, and human security are often used in reference with climate change, disease, food and water security, and migration. While sometimes used interchangeably, it is important to point out the significant differences between them, and to illustrate why the terms *transnational* and *transboundary* best capture the key analytical issue at the heart of this book: the cross-border aspect of these issues that generates the need for international cooperation as well as domestic policy action to effectively respond to them.

Some scholars have argued that the nature of new security threats requires new theories to explain them. We disagree. Our analysis is rooted in core concepts and theories from the discipline of political science, particularly international relations. From the realist perspective, our analysis focuses on national interests, threats, and security to illustrate not only the differences between traditional security threats and transnational challenges, but also the commonalities between them. Liberalism would posit that international institutions should be able to create incentives and opportunities for states to cooperate on common problems. Our study examines why this doesn't always happen and why, instead of seeing states act to provide public goods, we find states acting on behalf of private interests. As part of our discussion of international institutions, we also examine concepts of hybrid and fragmented governance. *Hybrid governance* refers to public-private partnerships that help promote collaborative solutions to collective action problems while *fragmented governance* recognizes that responding effectively to transnational security issues requires cooperation between many state and nonstate actors. It is not enough to understand how well or how poorly organizations help facilitate cooperation; it is also important to understand relationships among institutions, different layers of government, and nongovernment actors. Involvement of such a multiplicity of players can provide both opportunities and obstacles to cooperation, so we assess the reasons that lead states, institutions, and substate and private actors to work collaboratively on transnational security issues as well as the factors that produce conflict instead. We believe that a rigorous analysis of nontraditional security challenges in Southeast Asia that employs concepts in political science is the most fruitful way to address the key questions posed in this book.

National Interests and Security: Differentiating Traditional from Transnational Threats

All states have national interests that can be divided into three broad categories. First is the physical security of the state, which means protection against externally caused destruction of life and property within the territory of the state.[12] Second is its economic security, which means the economic prosperity and well-being of a country and its people. Third is the state's interest in the preservation of its values, civic culture, and forms of government.[13] All three of these national interests can be challenged by traditional military action undertaken by other states as well as by the transnational challenges discussed above. Migration, for example, can upset a state's physical security if the people crossing borders are combatants who carry the fighting across state boundaries or seek sanctuary from which to launch cross-border attacks. By imposing costs on receiving countries to house and feed people, migration threatens the economic security of states. Migration can also threaten a country's values and culture when migrants differ from local inhabitants in terms of race, ethnicity, or ideology and challenge the state's social stability, its governing ideology, or both.

In traditional security studies, threats are perceived as emanating from other states and take the form of military violence. Threats are therefore a product of an adversary's capabilities and its intentions. In the Asia Pacific region, North Korea is perceived as the region's most significant traditional security threat because its possession of nuclear weapons, combined with its growing ballistic missile technology, is raising its capacity at the same time that Kim Jong Un's aggressive rhetoric appears to signal that he fully intends to use his power projection capacity.

In this book, threats are considered from the perspective of the state being threatened.[14] Threats have a number of important characteristics that can help us understand the imperative for government leaders to combat them domestically and cooperate with other states to combat them collectively. First is the magnitude of the threat and, second, the probability that the threat will materialize. These two characteristics are often inversely related: the most severe threats can be those least likely to materialize, and leaders must therefore choose whether to respond to potent threats that may not ever occur or to more moderate ones that are more likely to happen. A third characteristic that influences policy is the imminence or timing of projected threats. Government leaders are more likely to respond to eventualities that are forecast to occur in the near term than to ones projected to materialize in the distant future. A fourth characteristic of threats is their tractability; that is, the degree to which they can be managed or eradicated effectively. Threats that are likely to materialize in the near term, and are easily dealt with, are more likely to be the object of government policy than those that are projected to occur in the future and for which there are no easy solutions.

Climate change, water and food scarcity, migration, and pandemic disease all pose significant challenges to Southeast Asian countries, but many of their negative consequences will happen in the longer term, which can reduce the incentives of states to respond to them now. Pandemics from EIDs have a low probability of occurring, but could pose extremely high levels of threat to a state's security and economic and social interests. Finally, many of the challenges posed by these transitional issues are fairly intractable since there are no easy solutions to global warming or threats posed from infectious diseases. All of these factors often make it difficult for governments to respond effectively to transboundary threats in the near term.

Transnational challenges do not pose the same degree of danger to all of the Southeast Asian states. Danger is the product of threat plus *vulnerability,* which can be defined as a susceptibility to injury or loss.[15] The degree of danger that each Southeast Asian state faces from nontraditional security challenges therefore is a function of the magnitude and probability of a threat and the state's vulnerability. Singapore may not be self-sufficient in food, but its wealth enables it to purchase food through trade on the international market. As long as Singapore has access to foreign-produced food, it is not vulnerable to food scarcity, even in times of rising prices or scarcity of basic staples such as rice. In contrast, states with significantly higher rates of poverty are much more vulnerable to food scarcity. Sharp price fluctuations in the rice market or decreases in the availability of basic commodities therefore may pose a danger to Cambodia or Myanmar, but not to Singapore.

The requirement that both threat and vulnerability be present for danger to exist points to two very different ways that states can respond to it. States who feel endangered may respond offensively by attempting to reduce or eliminate the threat itself. With regard to refugees such as the Rohingyas, Southeast Asian countries—in concert with other international actors—are attempting to pressure the government of Myanmar to treat the Rohingyas more humanely in an effort to reduce their incentives to leave Myanmar. If the Rohingyas stop fleeing their homeland, the challenges faced by Thailand and Malaysia will be eliminated. In contrast, states can respond defensively by attempting to mitigate the vulnerability that the threat seeks to exploit. For example, states threatened by sea-level rise produced by climate change can build levies to protect their cities from rising waters.

As Terry Deibel argues with respect to traditional security, "The relationship between interests, on the one hand, and threats and opportunities on the other, is one of the closest and most critical in all strategic thought."[16] We contend that studying the relationship between interests and threats is equally, if not more, important in the study of transitional issues. Recent decades have witnessed a broadening of the phenomena that threaten security beyond traditional military force as well as a deepening below the level of the nation-state to the actors who are negatively impacted. As James H. Mittleman argues,

> Nontraditional threats, including climate change, pandemics, transnational crime, and cross-border terror emanate from above and below the nation-state. Thus, there cannot be a neat separation between national and global security. Nor is there a sharp division between internal and external security. Sundry threats at home have extraterritorial dimensions . . . national security and global security, often regarded as counterpoints, are becoming a single stream.[17]

We agree that capturing the complex dynamics described by Mittleman are important. Too often, however, scholars have simply used adjectives to modify the term *security,* so that terms such as *nontraditional security* are defined in terms of what they are not rather than their core analytical features.

We believe that a close analysis of how the different transnational issues pose threats to the security, economic, and value interests of states and the people living within them best helps us understand the conditions under which governments adopt policies to respond to them at home and engage in efforts to promote strategic cooperation abroad. One of the key contributions of this book is to specify how the five transnational issues that it covers threaten the key interests of the state, its citizens, and the global community. The following analysis of the source and nature of the threats posed by transnational issues and the object being threatened is a step toward achieving this goal.

Comparing Threats and Security: Threats from What and Security for Whom?

When differentiating the transnational issues discussed in this book from traditional security, it is useful to compare them on two key dimensions that yield four distinct categories of security as illustrated in Figure 1.1: (1) the source of the threat, particularly whether it is military in nature; and (2) the object of the threat, particularly whether the state or the people living within it will suffer harm. Traditionally, national security has been defined in terms of external military threats to the state as in Cell 1 of Figure 1.1. Traditional security threats take the form of mobilized violence by one state against another state, as in the Vietnam War. This category of security is the core of realist security studies that equate national security with an absence of war.

In contrast, the source of threats can be military in nature but their object is not the state itself, but societies, groups, or individuals living within them as in Cell 3 of Figure 1.1. Intrastate conflicts, such as civil wars and ethnic conflict, that have grown increasingly common in recent years fall into this category. In the case of both traditional security and intrastate security, the nature of the threat is organized violence

Figure 1.1 Types of Security Threats

What Is the Source of the Security Threat?

		Military	Nonmilitary	
		Comprehensive Security		
Security for Whom?	States	Cell 1 National security (conventional realist approach to security studies)	Cell 2 Nontraditional security (e.g., environmental, health, and economic security)	Degree of Securitization
	Societies, Groups, and Individuals	Cell 3 Intrastate security (e.g., civil war, ethnic conflict, and genocide)	Cell 4 Human security (e.g., environmental and economic threats to the survival of societies, groups, and individuals)	

or military force. However, in contrast to realist notions of security that assume that the security of the state is coterminous with that of the individual, in Cell 3 we recognize that the physical territorial integrity of states and the security of the people living within them are different. In some cases, state rulers may be the principal threat to their citizens, as was the case in the Khmer Rouge in Cambodia and the current attempt by the Myanmar government to cleanse its state of the Rohingyas.

In contrast to military threats, in this book we focus on how nonmilitary or nontraditional threats such as climate change, disease, and water and food security impact national security. Here, the source of the threat is different, but the state remains the salient object of security as in Cell 2 of Figure 1.1. The term *comprehensive security* is often used in recognition that true national security requires protection from both the traditional military threats in Cell 1 and the nontraditional security issues in Cell 2. As Mutiah Alagappa observes, "Comprehensive security implied that the security of a state goes beyond (but does not exclude) the military to embrace the political, economic and social dimensions," all of which can be threatened by transnational security issues.[18] Comprehensive security has also been described in the following terms:

> Comprehensive security has two intertwined components: political security on the one hand (with its military, economic and social/humanitarian subcomponents): and environmental security on the other (with its protection-oriented and utilization-oriented subcomponents). To achieve comprehensive human security requires the satisfaction of both the political and the environmental sub-components—neither of these two major sub-components being either attainable or sustainable unless the other is satisfied as well.[19]

In Southeast Asia, the term *comprehensive security* has long been used by the region's postcolonial states and the Association of Southeast Asian Nations (ASEAN) in recognition that national security is not only a function of a lack of military conflict, but also socioeconomic development.[20] As the Indonesian government stated in 1973, comprehensive security "is an inward-looking concept, based on the proposition that national security lies not in military alliances but in self-reliance deriving from domestic factors such as economic and social development, political stability and a sense of nationalism."[21] These perceptions were shared by Malaysian prime minister Mahathir Mohamad who stated that "national security is inseparable from political stability, economic success and social harmony," and they have become embedded in ASEAN statements that recognize "that the concept of comprehensive security includes not only military aspects but also political, economic, social and other issues."[22]

When we look at issues of food and water security, climate change, the spread of infectious diseases, and migration, we believe (normatively) that states should view these problems as core elements of comprehensive security and engage with them seriously.

If state leaders perceive security in a fully comprehensive manner, then both the traditional and nontraditional issues in Cells 1 and 2 in Figure 1.1 will be perceived as national security threats. Despite the widespread use of the term *comprehensive security* in Southeast Asia, however, we find that states often fail to respond to transnational challenges as if they are threats to national security. Outside of efforts to promote food security when prices of staple goods soar or respond to an EID such as avian influenza, states have not acted as if these issues present challenges to state power and human well-being. A greater focus on comprehensive security can develop through shifting norms and priorities and this can be driven by international organizations.

Cell 4 of Figure 1.1 focuses on how transnational environmental, health, and other threats impact substate actors such as societies and individuals. Recent years have witnessed an increased focus on human security, or threats to the well-being of individuals and groups rather than the political unit to which they belong. In contrast to realist conceptions of security that assume that the security of the state is coterminous with the individual, the concept of human security is rooted in a traditional liberal concern with the rights and welfare of individuals. *Human security* is defined by the UN Development Programme (UNDP) in two parts: "first, safety from such chronic threats as hunger, disease and repression. And second, it means protection from sudden and harmful disruptions in the patterns of daily life—whether in homes, jobs, or in communities."[23] Food and water security, climate change, migration, and the spread of diseases are all problems that pose significant risks for human security across Southeast Asia.

The field of human security has expanded dramatically in recent years as scholars and policymakers have emphasized the welfare of ordinary people and formulated policies to enhance it. Indeed, a human security network among governments and nongovernmental organizations (NGOs) has arisen and many of the UN Sustainable Development Goals are focused on enhancing human security. Proponents of human security argue that the concept of security "has for too long been interpreted narrowly: as security of territory from external aggression, or as protection of nationalist interests in foreign policy or as a global security from the threat of nuclear holocaust. . . . Forgotten were the legitimate concerns of ordinary people who sought security in their daily lives."[24] Many scholars working on transnational issues in Southeast Asia have framed their studies in terms of human security. Mely Caballero-Anthony and Alistair D. B. Cook's groundbreaking book, *Non-traditional Security in Asia: Issues, Challenges, and Framework for Action,* offers a comprehensive discussion of the human security dimensions of climate change, the spread of disease, food security, and transnational issues. All of the examples we discuss here are threats to human security; however, they are not as clearly threats to national security. Government officials may perceive transnational issues as important problems and Caballero-Anthony and Cook's work offers a useful way of thinking about how problems become "securitized." "Securitization" involves several aspects: first, recognizing that an issue poses a significant threat to human well-being; second, according the problem a higher priority among resource imperatives; and, third, acting swiftly and with enough resources to address the problem. Caballero-Anthony and Cook find that, since these problems are transnational in nature, states have attempted to draw closer together to establish institutional and procedural arrangements to respond to these challenges. And they find that there is an increase in engagement between state and nonstate actors across local, national, and regional levels to address the problem.[25] We agree that transnational and nontraditional security threats do require these dynamics to take place; however, we find that cooperation is actually quite weak in addressing the problems and that cooperation among different actors within states happens when the more powerful actors (e.g., commercial rice exporters or commercial chicken farmers) want that cooperation to happen. Although countries in Southeast Asia talk a good game when it comes to diplomatic discussions about addressing nontraditional security issues, it is not always evident that governments see these threats as reaching the level of a security threat.

We recognize that the transnational issues examined in this book impact human security and we agree that efforts must be made to improve it. However, the definition of *human security* is broad, and some of the threats included in the definition, including "repression" and "sudden disruptions in the patterns of daily life" can emanate from states rather than the transnational issues at the heart of this book. In terms of our categories, therefore, transnational issues can indeed threaten human security, but so can other states and substate actors.

Furthermore, issues such as food scarcity and climate change do not impact all citizens of a country equally. Instead, people vary in their vulnerability to these threats. Wealthy individuals and groups will be much more immune from food scarcity than those living in poverty, just as the groups living in proximity to forests will be more vulnerable to the impact of deforestation and forest fires than those living far way. Therefore, one of the key questions we ask in this book is: In whose interest does the state act? This is a critical question for analyzing the domestic responses to transnational security issues as well as the prospects for international cooperation to address them.

In countries with governments that do value the security of their citizens, human and national security should be perceived as one and the same. The arrow in Figure 1.1 rising from Cell 4, human security, to Cell 3, nontraditional security, illustrates the extent to which governments equate human security with national security. If governments value the security of their citizens, then we would expect a high level of political will to meet transnational challenges and an investment in public resources to build the capacity to mount robust domestic responses to them. Countries adopting sound policies to respond to EIDs or climate change, in turn, will presumably be good partners for international cooperative efforts to address these challenges. Since transnational issues by definition cross borders, countries have an interest in what their neighbors and the global community more generally do to combat these dangers.

Not all governments act in the broad interests of their citizens. Since transnational issues may threaten some segments of society more profoundly than others, governments must choose in whose interests they will devote government resources and direct government policy. In the case of EIDs and avian influenza specifically, while all citizens may have an interest in freedom from infection, it is the poultry producers who must cull their flocks to reduce the risk of the disease that will suffer the economic costs. These private actors may therefore seek to influence government policy on this issue so as to reduce the costs to them. If such lobbying succeeds, then the poultry producers will have hijacked the national interest in favor of their private interests. If they are unable to influence government policy, the poultry producers may simply fail to comply with regulations that negatively impact them, forcing governments to adopt robust surveillance mechanisms.

Similarly, in the case of climate change, all citizens have an interest in reducing global warming, but those whose livelihoods depend on the forests may oppose efforts to limit deforestation. In such cases, governments must choose whether to act in the broad national (and regional or global) interests or in the interests of smaller segments of society. When efforts to combat a transnational issue require powerful interests to forgo profitable activities, they may be able to pressure governments to refrain from adopting comprehensive policies to respond to transnational issues or frustrate the implementation of those policies. When countries fail to adopt sound domestic policies in the face of transnational threats, they typically are poor partners for international collaborative efforts. To better explicate the conditions under which effective responses to transnational issues may arise, and how they are related to effective international cooperation, we now turn to a discussion of public and private goods.

Transnational Threats and Domestic Politics: Public Versus Private Goods

One of the central functions of government is to provide citizens with public goods that are necessary to promote broad national interests. *Public goods* are defined by two main characteristics: their consumption is nonrival and nonexcludable. The *nonrivalry* characteristic means that one's consumption of a public good does not detract from another person's consumption of the same good. The *nonexcludable* characteristic means that once the good is provided, it is open to consumption by everyone and people cannot be prevented from consuming it even if they did not contribute to the costs of providing it. A classic example of a pure public good is clean air: everyone can breathe it and there is no way to exclude someone from doing so. Another example of a public good is national defense. Public goods are often said to produce positive externalities, which are benefits that accrue to the entire community not simply those who pay for them.

In addition to pure public goods, there are many goods that have been termed *social goods* because, although they provide broad social benefits, they are not fully characterized by nonrivalry or nonexcludability. In the field of public health, for example, all citizens benefit when governments take steps to drain swamps to reduce the mosquito population and therefore reduce the risk of infection from mosquito-borne diseases such as malaria, dengue fever, or the Zika virus. An effort to reduce the mosquito population therefore is a social good, because it is something that all citizens consume. However, because the government must choose which swamps to drain, and in what order, there is a degree of rivalry because those living near swamps designated for draining presumably will benefit more than those living farther away.

In contrast, a country with limited resources may not be able to vaccinate all of its citizens to reduce their vulnerability to contagious diseases or to care for them once infected. In this case, access to vaccination is rivalrous because one's consumption negatively impacts the ability of others to consume that good. In contrast, some goods are rivalrous, but it is difficult to exclude individuals from consuming them in unsustainable amounts. For example, farmers may all benefit from large healthy grazing land but, since it is difficult to build fences around it, herders may overgraze on the land and thereby lead to its overexploitation. Such goods are often called *open pool resources,* and the collective dilemma here is often referred to as the *tragedy of the commons*: everyone would benefit from some limitations on consumption in the short term to ensure the sustainability of the land in the long term, but no private actor has an incentive or the authority to do so.

Public goods differ from *private goods,* which are those produced by individual actors for their own benefit. A farmer's rice is an example of a private good: a farmer can exclude others from consuming it and, once it has been consumed, it cannot be used again. In contrast to public goods from which positive externalities or benefits accrue, the actions of private actors may produce negative externalities, which are the unintended consequences or side effects of actions that are borne by others. A manufacturing plant, for example, may pollute the air or water, thereby reducing the quality of fresh air and clean water of others. A key function of governments, therefore, is to ensure that private actors refrain from imposing negative social costs on the broader community.

Ensuring that public goods are provided in sufficient quantities can be difficult because public goods are those that individual private actors or groups of private actors do not have incentives to provide for themselves. This is often referred to as the *collective action problem.* Everyone has an interest in efforts to eradicate mosquitoes or ensure clean air, but who pays the cost to ensure that these goods are provided? Since individuals, firms, and other private actors cannot be excluded from sharing in the benefits of clean air, they have an incentive to free ride on the willingness of others to provide them. If everyone free rides, then the goods will not be provided in sufficient quantities, even though it is in everyone's interest that this be done. A key responsibility of governments, therefore, is to act in the broad national or public interest and help ensure that collective action problems are overcome.[26]

Doing so, however, requires that governments define the national interest in ways consistent with the public interest, as in the classic liberal formulation where the national security interest is simply the sum of the individual interests of its citizens. It also requires that governments have the capacity to provide public goods. On key transnational issues in many Southeast Asian states, however, these two conditions may not be met.

In wealthy democratic states, governments have access to resources to provide public goods and there is normally a lively political discussion about how much should be spent on public goods and how the costs of these goods should be apportioned.[27] In developing states, the government's willingness and ability to overcome the collective action problem to provide these goods to citizens may be more problematic. Governments often have conflicting priorities. In many countries, powerful private actors may lobby the government to act not in the broad national interest, but in their private interests. Manufacturing firms may oppose government efforts to limit and regulate the pollution-producing activities from which they profit. Governments dependent on powerful private actors for political support may choose to reward their supporters rather than adopt policies consistent with the broad national interest. Similarly, governments in many developing countries have fewer resources to spend on public goods and governments may have weaker capacity to deliver or implement public works projects. In other words, developing countries are often less successful at solving collective action problems.

Food and water security, climate change, the spread of new and deadly diseases, and migration flows all require us to think about public goods in two ways. First, it is the role of government to provide for the well-being of its citizens. This includes the ability of people to be able to feed themselves and their families. It also includes access to water for personal use such as drinking water, sanitation, and agriculture, since in many developing countries rural populations rely on growing crops for at least part of the household's dietary needs. It includes the responsibility of the state to respond to natural disasters such as flooding, drought, and pandemic disease. Many of these needs represent policy choices relating to the provision of public goods where governments have the ability to take action on their own to meet citizens' needs domestically.

Many of the transnational issues discussed in this book, however, create problems that states cannot solve on their own. Governments cannot ensure the water supply for their own citizens living along the Mekong River if a different country upstream builds a dam that changes the water flow, silt, and salinity, which adversely affect the amount and quality of water available in their country. Governments cannot ensure access to staple food crops such as rice if they are beholden to export conditions from other countries; neither can they always protect citizens from the effects of environmental degradation taking place in other countries, or from the spread of a disease outbreak that begins outside their borders.

Just as the actions of private actors may produce negative externalities or public "bads" that require government action at the domestic level, actions taken by public or private sector actors in one country can create negative externalities for other countries. Thus, states need to cooperate with each other to be able to provide these larger public goods and overcome the collective action problem at the international level. International organizations (in this book we use the terms *international organization* and *international institution* interchangeably), have been created, in part, to help provide public goods that states cannot necessarily provide on their own. Thus, the problems discussed in this book need to be understood as part of a larger discussion about the role of international or regional organizations to solve these kinds of collective action problems and to help states improve their ability to provide public goods to their own citizens.

International Institutions, Global Governance, and Cooperation

Questions about cooperation are not new in international relations, but the literature on Southeast Asia and nontraditional security threats is marked by an absence of comprehensive and theoretically grounded work. Addressing transnational threats requires states to overcome the collective action problem which holds that, even when states share common interests in collaboration, they will attempt to free ride on the efforts of others to avoid paying the costs. States therefore have mixed interests, or incentives to cooperate and not cooperate. All states can benefit from *cooperation,* which is defined as "goal directed behavior that entails mutual policy adjustments so that all sides end up better off than they would otherwise be."[28] The key question is how to structure incentives so that states perceive that their interests are better served by making the policy adjustments necessary to capture those gains from cooperation.

Institutions play an important role in promoting cooperation between states. The literature on international institutions is voluminous due to their proliferation in recent years.[29] For the purpose of analyzing transnational security issues, in this section we focus on identifying the different types of international institutions, the mechanisms by which they help promote cooperation, and how they work with other actors in global governance to overcome the collective action problem.

There is no common definition of institution in the literature.[30] Traditionally, international organizations have been conceived as formal institutions whose members are states, which are sometimes referred to as international intergovernmental organizations. An *international organization* can be defined as a formal continuous structure founded by an authoritative instrument of agreement between member states.[31] Elaborating on this definition, the requirement that an international institution be formal and have continuous structures means that institutions must have a headquarters or international secretariat responsible to the organization, a formal set of rules, decisionmaking procedures, and consultative organizations that meet regularly. This distinguishes institutions from conferences or meetings seeking one-time solutions to immediate problems such as those that produced the agreement between the United States and Japan to release rice at the height of the 2007–2008 rice shortage, which helped resolve the food security crisis at that time.

Authoritative instruments of agreement can take a number of forms, but they are often international treaties ratified by states. Governments of states voluntarily join, contribute financing, and make decisions within the institution. International organizations are identifiable because their purpose, structure, rules, and decisionmaking procedures are clearly specified in a charter or treaty. International institutions can be categorized by the openness of their rule of membership as well as the scope of their purpose. The United Nations is a universal organization because all states can join. In contrast, many organizations restrict membership based on geographic membership, wealth, or other criteria. Membership in ASEAN is open only to the states in Southeast Asia, although it has created offshoots such as the ASEAN Regional Forum (ARF) and the East Asia Summit (EAS) to engage outside actors.

Institutions can also be categorized according to their mission, or the specific issue area they were created to address. Both the UN and ASEAN are comprehensive multipurpose organizations with broad mandates that were originally created to help promote peace. In contrast, the WHO is responsible for addressing international health issues while the Food and Agriculture Organization (FAO) is dedicated to helping ensure sufficient food supplies to combat hunger and food insecurity for a growing global population. Both the WHO and the FAO were created under the auspices of the UN and function as UN specialized agencies, illustrating how existing institutions can help create new ones as well as the recognition that helping to resolve specific transnational issues such as health and food security often requires more focused international collaboration than multipurpose efforts can provide.

Institutions help states cooperate in a number of ways. First, by "extending the shadow of the future," or ensuring that states will interact with one another over time rather than in single transactions, they create prospects for future cooperation and, thereby, help to promote reciprocity and expectations of increasing potential gains from cooperation. They also reward states that develop a reputation for compliance with institutional agreements and penalize those that break them. Second, institutions can increase the amount of information available to states participating in international agreements by creating a framework for the sharing of information. Transparency about the preferences of other actors and more accurate information about the transnational issue at hand makes it easier to identify the benefits of cooperation. In addition, increased information and transparency also make better monitoring possible. Third, institutions can lower transaction costs. By helping to disseminate information, negotiate agreements, and monitor compliance with their rules, institutions enhance the efficiency of cooperation, which makes it more attractive for states.

The extent to which institutions can compel states to act in accordance with the rules and norms of the organization varies tremendously. Countries in the European Union (EU) have ceded sovereignty to the EU in many issue areas, enabling the EU to enforce its laws and rules and compel its member states to uphold their commitments. In contrast, Southeast Asian states guard their sovereignty jealously and have not ceded any supranational authority to ASEAN. ASEAN, therefore, is a much weaker regional organization than the EU. It lacks an effective compliance mechanism which, as this book illustrates, has hampered effective cooperation on transnational security issues in Southeast Asia.

Countries therefore cooperate more successfully when there are robust international or regional institutions to lower transaction costs, create incentives for states to cooperate, and impose penalties for noncompliance. And institutions are more able to do this if there is strong leadership within them. Ideas about what best promotes leadership in international organizations has changed over time. Hegemonic stability theory argues that a single predominant power or hegemon may have the capacity for, interest in, and commitment to a set of ordering principles to shoulder most of the costs involved in solving an international problem.[32] Viewing international economic and financial stability as a global public good, Charles P. Kindleberger argues that a hegemon was necessary to ensure financial stability because only a predominant power like Great Britain or the United States had the economic resources, commitment to a liberal international order, and willingness to act as a lender of last resort during financial crises.[33] Similarly, many scholars contend that US hegemony was necessary for the creation of many post–World War II institutions such as the World Bank, International Monetary Fund (IMF), and World Trade Organization (WTO) that promote economic cooperation.

Today, globalization has led to the diffusion of power in the international system as evidenced by the emergence of new powers like China and the relative decline of the United States. Thus, Kindleberger and others argue that institutions could survive after hegemony and the cooperation they foster could persist in the absence of a predominant power.[34] As long as countries with an interest in supporting the institution's mission and the capacity to do so work together, cooperation could continue in a more decentralized manner. Consequently, effective institutional leadership is typically a result of the active participation by countries with an interest in collaborative solutions, the capacity to adopt effective policies, and a stake in the issue sufficiently large that their actions can alter outcomes.

In Southeast Asia, Indonesia is the region's predominant power and, consistent with the literature on hegemons and pivotal states, it has played an important leadership role in ASEAN on traditional security and diplomatic issues promoting regional peace and stability.[35] On transnational issues such as climate change, migration, and disease, however, Indonesia is often perceived by its neighbors as a source of threat rather than a source of leadership. The haze caused by forest fires threatens regional health security, Indonesian migrant workers in Malaysia are viewed as threats to social stability, and its poor responses to outbreaks of avian influenza threaten regional health security. ASEAN, like the UN, is an intergovernmental organization that has spawned a host of working groups, committees, and expert panels to discuss transnational challenges, develop best practices, and at times outline regional treaties and agreements on them. Indonesia, however, signed the ASEAN Agreement on Transboundary Haze Pollution a decade after other members, and halted some cooperation with the WHO during the height of the avian influenza outbreak. Indonesia has taken a leadership role in pushing for an ASEAN agreement on migrant workers, but its efforts have been opposed by other members. In Southeast Asia, the disconnect between leadership in political organizations like ASEAN and that on transnational security issues hampers regional cooperation.

Changes in the global distribution of power, combined with the rise of new global challenges such as the transnational security issues analyzed here, have meant that promoting cooperation has required states and institutions to collaborate with a broader array of actors to solve global challenges. A key role of international institutions therefore is to foster cooperation among all relevant actors in global governance. Global governance is fundamentally about problem-solving arrangements and activities that states and other actors engage in to address common problems. The Commission on Global Governance defines *governance* as "the sum of the many ways individuals and institutions, public and private, manage their common affairs. It is a continuing process through which conflicting or diverse interests may be accommodated and cooperative action may be taken. It includes formal . . . as well as informal arrangements that people and institutions have agreed to or perceive to be in their interest."[36]

International governance is not the same as international government. Domestically, governments have the ability to coerce compliance. Governments have police and courts to enforce laws made by a legislative body. In contrast, global governance, as Emmanuel Adler and Steven

Bernstein note, decouples coercive force from legitimate rule making. Institutions and networks may make rules and create norms of behavior, but without the same enforcement power of a state.[37] A key challenge in addressing transnational threats such as climate change, food and water security, disease, and migration is finding ways to ensure that states and other actors comply with policies designed to redress them.

Governance therefore requires participation by, and cooperation between, states to meet collective challenges. Effective global or regional governance is predicated on changes in state behavior. This is often a difficult and contested occurrence because it may mean that a state not only must change its own behavior, but also must attempt to enforce changes on private actors. When responding to transnational threats, states often face the dilemma that the most effective collective responses will require them to give up a measure of sovereignty. State sovereignty is an enduring and powerful norm in international relations. As noted earlier, European countries have ceded significant elements of sovereignty to the EU, but Asian states have not been willing to do so. Organizations in Asia such as ASEAN and its many offshoots are profoundly weaker institutions than the EU because Asian states have agreed to participate in these organizations with the express understanding that their sovereignty will be respected, not compromised. Indeed, respect for state sovereignty and noninterference in the domestic affairs of member states are cardinal ASEAN rules. These norms may secure the participation of regional states in organizations like ASEAN but they ultimately may conflict with attempts to create greater governance mechanisms, particularly those requiring compliance mechanisms.[38]

States may still be the primary actors but promoting collaborative solutions toward transnational security issues increasingly also involves nonstate actors in what scholars term *hybrid governance.* Hybrid governance occurs when "non-state actors co-govern along with state actors for the provision of collective goods, and adopt governance functions that have formally been the sole authority of sovereign nation-states."[39] The increasing participation of nongovernmental actors at the global level is a reflection of the ongoing reconfiguration of world politics as well as the rise of new transnational issues such as climate change. International nongovernmental organizations (INGOs) are key actors in many of the transnational issues analyzed in this book. INGOs are private nonprofit organizations engaged in a variety of activities. Like their state counterparts, they can have a comprehensive multipurpose agenda or a more specific one. They help promote cooperation by defining goals, providing information, and giving advice. They also can pressure governments through direct or indirect lobbying.

Private for-profit actors such as transnational corporations (TNCs), sometimes referred to as multinational corporations (MNCs), also play a role in governance. TNCs are business organizations that have subsidiaries in at least two states and engage in transnational production or extraction activities involving the movement of goods and services across state boundaries. Global pharmaceutical companies that have the technology, funding, and production capacity to produce vaccines can play a critical role in helping to respond to outbreaks of EIDs if they help to develop and produce the necessary vaccines in a timely fashion and distribute them to the countries and people who are most negatively impacted by the EIDs. In contrast, they can engage in beggar-thy-neighbor policies if they demand high profits or otherwise impede the production and distribution of medication to those most directly affected by an outbreak of an EID. Often, states and international organizations work with INGOs, such as the International Red Cross and Red Crescent Movement and Doctors Without Borders, to help meet the challenge of responding to EIDs.[40]

Effective responses to all of the transnational issues discussed in this book require cooperation between different types of actors—international institutions, states, NGOs, and private actors—at the global, regional, national, and substate levels. Efficacy therefore requires both horizontal and vertical collaboration. In the case of climate change, for example, horizontal collaboration at the global level occurs when states come together under the auspices of the UN Framework Convention on Climate Change (UNFCC) to be apprised of the latest scientific data by the Intergovernmental Panel on Climate Change (IPCC), a group of scientists who provide policymakers with assessments of climate change, its impacts, and its future risks. Based on projections of the negative impacts of rising temperatures, a consensus was reached to aim to limit the rise in global warming to 2 degrees C over the twenty-first century, a goal enshrined in the 2015 Paris Agreement. To induce developing countries to ratify the agreement, which requires all states to submit national climate action plans, richer countries agreed to provide funding to help them implement the action plans. Many nonstate actors were also involved in these negotiations, ranging from INGOs such as the World Wildlife Fund (WWF) and Rainforest Alliance to hundreds of TNCs, many of whom made commitments to lower their carbon footprint.

Once global agreements to combat climate change are made, they must be implemented effectively on the ground and this requires coordination vertically across multiple layers of government. Implementing Indonesia's commitment under the Paris Agreement not only requires national-level policies such as a moratorium on the conversion of forests to agriculture and mining use, but also that this moratorium be implemented by provincial and local-level officials on the ground. Reformist local officials often work with NGOs to monitor activities on forestland by powerful actors such as logging and mining companies that have an interest in evading the moratorium. These NGOs, in turn, often receive training, funding, and equipment such as drones from INGOs with which they are affiliated.

The term *fragmented governance* is often used in recognition of the fact that effective solutions to transnational security threats require vertical and horizontal cooperation between a multiplicity of state and nonstate actors.[41] Fragmented governance refers to the diverse, sometimes overlapping or competing institutions, agreements, and agencies tasked with resolving a specific problem.[42] Continuing with the climate change example, Indonesia signed a bilateral memorandum with Norway that will provide up to $1 billion for verifiable reductions in GHGs that produce global warming, it is a member of the multilateral Forest Carbon Partnership Facility, and it recently merged its Ministry of the Environment with the Ministry of Forestry in effort to enhance efficiency and coordination between these two ministries that often had conflicting goals. All of these agreements, institutions, and agencies, and the processes they oversee, must then be coordinated or networked to be effective. Fragmented and networked governance, therefore, are two sides to the same coin.[43]

Referring to these dynamics as fragmented implies looking at smaller units of analysis or parts of a whole. Attempting to ensure that all commitments Indonesia has made to combat climate change and that the actions taken by the government, NGOs, and private actors work together to reduce GHGs illustrate the need for these activities to be networked or coordinated. The problems that we examine in this book are ripe for looking at fragmented forms of governance and, in each chapter, we do examine how different actors and sets of actors work to cooperate or stymie cooperation on shared problems.[44]

There is no consensus on the consequences of fragmentation for global governance. Fragmentation is a matter of degree and varies among issue areas. In some instances there are positive outcomes, in others negative ones, and in some we find mixed results. Frank Biermann, Philipp Pattberg, Harro van Asselt, and Fariborz Zelli develop a typology of three types of fragmentation: (1) synergistic (when different institutions and actors have overlapping, complementary, and coordinated approaches to solving a problem); (2) cooperative (when institutions and actors work well and cooperate on shared problems); and (3) conflictive (when institutions and actors are hardly connected and have different decisionmaking procedures, conflicting principles, norms, and rules, and different memberships and priorities).[45] This typology helps differentiate between positive and negative dynamics and their sources.

Examining the impacts of fragmented governance on cooperative efforts on river basins, a key to enhancing water security for riparian states, Neda A. Zawahri and Sara McLaughlin Mitchell find that different factors influence the rise of bilateral versus multilateral agreements on shared (multilateral) river basins.[46] States are more likely to sign bilateral agreements on shared water, yet the basins are multilateral and an integrated management approached is advocated by environmentalists, international organization theorists, engineers, and water experts. Zawahri and Mitchell also find that treaty types are a product of state interests, transaction costs, and distribution of power of upstream and downstream states.[47] In a work on cooperation over the Ganges River, the authors find that "the riparians' failure to coordinate their development of the river through a multilateral effort has minimized the gains from cooperation, contributed to inefficient resource use, and resulted in environmental degradation of the basin."[48] We reach the same conclusion regarding the impact of fragmentation on efforts to govern the Mekong: there are multilateral frameworks in place for cooperation over the Mekong, but little cooperation exists on water rights and usage. Instead, we find that fragmentation produces conflictive outcomes rather than cooperative ones on the issue of water security.

The strength of the governance networks in each issue area varies tremendously. International organizations, as discussed above, do not always succeed in getting states to cooperate on common problems. Organizations function most effectively when they have strong leadership. Strong leadership, in turn, depends largely on whether the interests of powerful actors are affected by these issues and whether they are willing to devote the time, diplomatic effort, and material resources to formulating solutions to the challenges posed by transnational threats. The extent to which the interests of powerful states are impacted, in turn, depends in part on the extent to which these issues threaten global public goods.

Of the five transnational issues covered here, global warming produced by climate change and the spread of infectious disease are the threats that most clearly impact the interests of powerful actors. Global warming is changing weather patterns throughout the world, producing sea-level rise, changes in rainfall, and extreme weather events that impact all countries, albeit in different ways. Similarly, diseases spread indiscriminately across borders. Wealthy countries with healthy populations, well-developed heath care systems, and an indigenous pharmaceutical-producing capacity will be better able to respond to threats of pandemic disease, but they are not immune from the threat itself as outbreaks illustrate. The environment and global public health are two issues that most closely approximate global public goods, and we would therefore expect to see robust global governance to promote efforts to overcome the collective action problem and manage them.

Indeed, it is the case that institutions and governance at the global level are strongest in these areas. The UN, as discussed above, is a broad multipurpose organization that has developed many specialized institutions over the decades to help coordinate global governance on all of the issues discussed in this book. Under UN auspices, the UNFCCC has coordinated the response to climate change, which involves not only states but panels of scientific experts, NGOs, and private actors. Similarly, the WHO has promulgated international health protocols that outline the responsibilities of states to report cases of infectious diseases to help monitor their spread and it coordinates activities of national health agencies, private health care providers, and NGOs to help prevent, and if necessary respond to, outbreaks of infectious disease.

There are some international treaties, conventions, and UN agencies responsible for governance on migration and water and food security, but global governance in these areas is not as robust as in climate change and disease because these issues have localized effects rather than standardized ones across the world. With regard to migration, the UN has conventions on refugees and migrant workers, a UN High Commissioner for Refugees (UNHCR), and a well-established set of protocols governing the treatment of refugees. While migration is rising globally, it does not impact all countries equally. Refugees fleeing conflicts and economic migrants seeking opportunities disproportionately impact neighboring countries. As the UNHCR global trend report illustrates, while the influx of migration to Europe from people fleeing conflict in the Middle East and North Africa has garnered global attention, Syria's neighbors harbor far more refugees. The UNHCR has complex rules governing the rights of refugees and a large network of states, NGOs, and private actors assisting them, but the global regime is breaking down in part due to a growing unwillingness of states to offer permanent resettlement to refugees when it is unsafe for them to return to their homelands. Germany took a leadership role on migration by offering to take in 1 million refugees in 2015, but this generated a backlash at home and abroad. Labor migration, in contrast, is typically governed by bilateral agreements between sending and receiving states, with the international conventions playing a small role. Governance on migration is therefore extremely fragmented.

Threats to food and water security pose the most localized impacts and, therefore, have the least developed global governance structures.

The UN created the FAO, but efforts to grapple with the threats posed by sharp shifts in the supply and price of staple foods to food insecure countries butt up against the market mechanism and the profits it generates for countries and private actors. With respect to water, the world's oceans are indeed a global public good, and therefore the UN Convention on the Law of the Sea (UNCLOS) is a well-developed body of international laws to govern this part of the global commons. International conventions on transboundary rivers attempt to lay out the rights and responsibilities of upstream and downstream countries, but regional mechanisms to govern water are often more robust. In the case of the Mekong, the Mekong River Commission (MRC) has existed for decades, but its capacity to govern the Mekong has diminished over time as upstream counties build dams that negatively impact downstream countries and China refuses to join. China, as the region's hegemon or predominant power, has the capacity to play a leadership role in forging a cooperative solution to ensure fair distribution of the Mekong's water, but it instead has chosen to promote its own water security by damming the Mekong.

Too often national policies on transboundary security issues, such as the damming of the Mekong by upstream countries, exemplify beggar–thy-neighbor policies where countries enact measures to protect their own producers, companies, and citizens at the expense of those outside their borders. The larger effect of these policies can sometimes be a worsening of conditions regionally. While these actions are at times understandable, they illustrate a fundamental problem of international relations. What one country does impacts other countries. For example, when a food exporter decides to reduce or stop exporting that product, it can have dire consequences elsewhere for those who rely on importing that food commodity. Similarly, the burning of forests to clear land for agriculture or commercial plantations not only exacerbates global warming, but pollution from the fires can negatively impact the health and well-being of its neighbors. Cooperation often could make countries better off, but states find it difficult. Cooperation is hindered by a lack of leadership, imperfect information, a lack of regular mechanisms for designing agreements, and variation in the level of interest from states in cooperating on any given issue.

One of the reasons we do not see greater international cooperation is that powerful groups within countries have been able to assert and protect their narrow interests, which are often at odds with efforts for greater cooperation. When powerful political or economic actors favor protectionist policies, or noncooperative behavior, then it is more likely that cooperative efforts will fail. In contrast, when interest groups stand to gain from international cooperation on issues ranging from trade agreements, environmental agreements, transboundary cooperation, or research on emerging infectious diseases, we are more likely to see implementation of policies that will lead to cooperation.[49]

The Plan of the Book

In this book, we analyze each of the five transnational security issues discussed earlier to assess the extent to which there are common or competing interests among the countries and the extent to which there is cooperation or conflict. Each chapter focuses on a key topic within the broader transnational issue and outlines the relevant international, regional, and bilateral institutions, conventions, or protocols that attempt to govern the issue. At the same time, we examine the domestic policy context to assess whether government policy is made in the broad national interests or in more particular ones. We also discuss whether powerful political and economic actors have an interest in cooperation or conflict, and assess the capacity of states to respond effectively to these transnational challenges.

Chapter 2 examines the problem of climate change. In Southeast Asia, deforestation is the largest contributing factor for the region's GHG emissions and Indonesia is by far the largest regional emitter. In this chapter, we therefore focus on Indonesia's efforts to reduce its rates of deforestation and the related issue of preventing and responding to the transboundary haze caused by forest fires on peatland. We begin by examining Indonesia's role in the UNFCCC to illustrate that it has played a key mediatory role between developed and developing countries in climate change diplomacy and has made significant commitments under the UNFCCC. Indonesia's ability to meet its international commitments, however, has been hampered by a lack of technical capacity, poor governance in the forestry and land sectors, and efforts by powerful private actors who benefit from the current system to thwart efforts to improve governance. We analyze these issues by focusing on the government's efforts to implement Reducing Emissions from Deforestation and Forest Degradation (REDD and REDD+), restructure the chaotic governance structure, and enact a strong moratorium on forest conversion. We illustrate that despite significant international support and the threat climate change poses to Indonesia's broad national interest, powerful private actors have been able to stymie effective action, as was made abundantly clear by the 2015 haze. Fragmented governance has been a significant obstacle to better implementation of bilateral and multilateral agreements to address the problem.

Chapter 2 also analyzes efforts to halt the transboundary haze that threatens the public health of Indonesia's own citizens as well as those in neighboring countries, particularly Singapore and Malaysia. The haze is an extreme case of negative externalities, in which the actions of private actors in Indonesia impose tremendous health, social, and economic costs on neighboring countries. In this chapter, we examine the regional conventions created to address the haze, such as the ASEAN Agreement on Transboundary Haze Pollution, which have failed to have an impact on reducing the fires or responding effectively to them. We discuss Singapore's efforts to hold Indonesian plantation owners accountable for their actions, which has triggered only conflict, rather than cooperation, among Southeast Asian states. We then discuss the new and potentially significant measures adopted by the Joko Widodo (aka Jokowi) administration in 2016 in response to the recent haze that, if implemented, could be potential game changers in Indonesian climate change efforts.

Chapter 3 turns our attention to problems of food security. In the chapter, we focus on the rice trade in Southeast Asia to demonstrate how access to food requires cross-boundary cooperation. Rice is a critical crop in Southeast Asia, and an analysis of the problems and possible cooperation over the rice trade demonstrates the larger questions and problems inherent to food security more generally. Malaysia, Singapore, and the Philippines are rice importers.[50] For Malaysia and the Philippines in particular, rice security is a matter of national security. Both countries are instituting national policies to bolster national production,[51] but they are still reliant on rice imports to fulfill consumer demand. Thailand, Vietnam, and Cambodia are rice exporters. In all three countries, governments are seeking to protect both their own producers from price fluctuations and their own citizens from potential global shortages. In theory, a region characterized by some of the world's largest rice importers and exporters would seem ripe for collaborate efforts to promote food security.

Partly in response to the sharp increase in international food prices in 2007–2008, leaders in Southeast Asia pledged to embrace food security as a matter of permanent and high priority. They adopted a "Statement on Food Security in the ASEAN Region," which commits, among other actions, to the implementation of the ASEAN Integrated Food Security Framework (AIFS) and the Strategic Plan of Action on Food Security in the ASEAN Region (SPA-FS) (2009–2013). This was a five-year strategic plan to assure long-term food security in the region through increased cooperation and mutual help.[52] These measures could move the region toward greater cooperation on the rice trade, but few steps have been undertaken to operationalize these mechanisms due to the competing interests of rice importers and rice exporters.

Exporters Vietnam and Thailand are contemplating greater cooperation on creating a rice cartel under the umbrella of a regional federation for rice among ASEAN rice-producing nations. The hope is that rice exporters will be able to command higher prices for their farmers if they work together as a group to set prices and control supply.[53] Creating a cartel-like situation would benefit suppliers over consumers because of their ability to control prices. Importing countries, such as Malaysia, the Philippines, and Singapore, would like to see as open, transparent, and competitive a market as possible for rice. Since the priorities and preferences are so divergent for importers and exporters, it is hard to envision a situation where greater coordination of rice trade policies will be possible. Cooperation and competition on the rice trade reflect the phenomenon of networked governance. We do see a patchwork of efforts from private and public actors to tackle pieces of this problem. We see bilateral agreements, powerful interest group efforts, some regional cooperation on sharing of information and stockpiling some rice for emergencies, but as of yet no comprehensive collective effort to make sure that there is no repeat of the rice crisis. No comprehensive set of rules and procedures have been developed for exporters. In this case, the fragmented nature of interests and efforts has stymied prospects for greater regional cooperation on food security.

The focus of Chapter 4 on water security moves beyond failures to provide citizens with public goods such as clean water and sanitation delivery, and examines regional efforts to cooperate on shared resources like the Mekong River. Shared by China, Myanmar, Laos, Cambodia, Thailand, and Vietnam, the Mekong is one of the world's greatest waterways. It is used by 60 million citizens (just along the lower Mekong) as their main source of water for drinking, bathing, transportation, and economic livelihoods in the form of water for irrigating crops and for industry. It is increasingly also being used as a source of hydropower with the construction of dam projects in several of the upstream riparian states. The damming of the Mekong by upstream countries such as China and Laos is negatively impacting downstream countries in classic beggar-thy-neighbor fashion.

Over the years, a number of international organizations have been created to attempt to govern the Mekong. In 1995, the Mekong River Commission was created to help provide oversight and cooperation on issues ranging from drought and flood management, to agricultural and irrigation needs, to fisheries and the environment, and to hydropower and navigation. At its core, MRC is charged with ensuring the reasonable and equitable utilization of the Mekong's water.[54] While MRC had some success in studying, mapping, and surveying the river's resources, it is currently facing significant challenges.

First, the MRC members are Thailand, Cambodia, Laos, and Vietnam, with China and Myanmar participating as observers. No organization that fails to include China, the source of the Mekong and hence the country with the largest degree of control over its waters, could alleviate key collective action problems. More critically, dam building by China and Laos on the upper reaches of the Mekong have significantly changed the river's flow and silt. At the 2014 ASEAN-China summit meeting, Chinese premier Li Keqiang proposed the establishment of the Lancang-Mekong River Cooperation Mechanism (LMRC). The organization came into existence quickly, and its stated aims are to promote cooperation and consensus on political security, economics, and sustainable development surrounding the river and its use. However, China has resisted the idea of codifying rules for water use, dam construction, and the like. China did release water in 2016 to alleviate some of the worst effects of the drought in Southeast Asia, but this was a unilateral action and downstream countries have no assurances that water will be forthcoming in the future. Governance of the Mekong has been characterized much more by conflict than cooperation. This may be by design. As work on fragmented governance suggests, when a great power like China prefers to negotiate bilateral deals and the organizations they spearhead tend to be focused on China's trade preferences and weak on enforcement mechanisms and provision of public goods, we are unlikely to see multilateral cooperative results. Biermann, Pattberg, van Asselt, and Zelli's description of conflict arising from disharmonious organizations and priorities match Chinese efforts on the Mekong.[55] China has not proposed any agreements on shared water rights and has not offered any codified assurances to downstream states about its intentions for the Mekong.

Chapter 5 looks at the issue of disease and health security through an analysis of efforts to address the emergence of avian influenza in Southeast Asia. Outbreaks in China, Vietnam, Cambodia, Thailand, and Indonesia have triggered concerns about a global health crisis that could start in the region. H5N1 is largely spread through bird flocks, and it is carried and transmitted by both wild birds and livestock flocks. Humans almost always contract the virus from handling sick or infected birds. Effective protocols for preventing disease outbreaks in humans include: quarantining both new birds when introducing them to existing flocks, and quarantining and culling infected birds; sanitary rearing and handling practices, as well as guidelines for safe food handling techniques and preventing consumption of sick birds. While some countries in the region have been effective at instituting and applying such measures (e.g., Malaysia and China have seen rates of transmission fall since 2003), Cambodia, Indonesia, and Vietnam struggled to fully address outbreaks of larger numbers of human cases of H5N1.[56]

Reasons for the difficulty in either eradicating the incidence of H5N1 or reducing its frequency are numerous. First, in countries that are not well-off, it is prohibitively expensive to adequately pay farmers to cull sick flocks and to pay them well enough to incentivize them to carry out accepted international protocols for cutting down the transmission of the disease. Second, corruption and local-level political dynamics make it harder to successfully implement national programs in places where local officials may have significant power as in Indonesia and Cambodia. Third, there are domestic political factors such as bureaucratic disagreements on policy and jurisdiction as well as the potential political costs to elected leaders of advocating unpopular policies like flock culling ahead of elections.[57] In short, failures of political will and capacity hamper effective responses. What we find in looking at how states have addressed H5N1 are examples both of successful networked governance architecture and of failures arising from fragmentation. In Vietnam and Thailand, coordinated efforts and shared priorities from national and local government bodies and nongovernmental actors like poultry farmers and agribusiness interests have produced successful results. In Cambodia and Indonesia, we find conflicting interests and poor coordination among such groups.

Most international work on EIDs is conducted through the auspices of the WHO. The US Centers for Disease Control (CDC) in Atlanta works closely with the WHO around the world in helping to identify and track global pandemics. ASEAN is also working to address EIDs. There are a myriad of existing ASEAN bodies already in operation to address the threat of pandemics, but ASEAN's difficulty in being able to do so lies with the wide capacity gaps in health systems among member states,[58] and perhaps the lack of willingness to create a regional fund to help poorer countries follow accepted protocols for culling diseased flocks. Chapter 5 demonstrates that ASEAN seems to be making the most progress on the least controversial of these measures such as establishing websites to share information on EIDs (maintained by Indonesia). Countries have not always been forthcoming in sharing information, and compliance with best practices has been spotty at best.

Chapter 6 examines the issue of migration through an analysis of Indonesian migrant workers in Malaysia and the Rohingya refugees. Labor migration should be a source of regional cooperation as labor-scarce countries (e.g., Malaysia and Singapore) import workers from countries with surplus labor (e.g., Indonesia and the Philippines). However, in this chapter we find that labor migration triggers conflict when it shifts from being perceived as an economic issue that produces gains for both countries involved to one that becomes defined as a political and security issue in which the interests of sending and receiving states are viewed as threats to one another. Indeed, the issue of migrant workers has become so politically contentious that Indonesia and Malaysia have taken steps to stop it: Indonesia by issuing a ban on female citizens migrating to work as domestic workers and Malaysia by issuing decrees to ban its employers from hiring migrant workers. We find that better domestic governance of migrant workers in both countries could help ameliorate conflict in the relationship, but neither side has been willing to take on powerful vested interests to do so.

Chapter 6 also examines the Myanmar government's persecution of the Rohingyas, which has produced an exodus of people and triggered conflict in Southeast Asia. According to the UNHRC, in 2015—a year which witnessed a record number of stateless people, refugees, and displaced persons—Myanmar ranked as the eighth-largest source country of refugees, with 451,800.[59] The Rohingyas are viewed as refugees, and should therefore be granted temporary refuge and protection by receiving states under the Convention Related to the Status of Refugees, also known as the UN Refugee Convention. Instead, many Southeast Asian states have adopted beggar-thy-neighbor policies of pushing the Rohingyas back to sea and attempting to avoid paying the social and economic costs of hosting them. The Rohingya issue, therefore, is triggering conflict not only between Myanmar and its neighbors, but also among the Southeast Asian states to which the Rohingyas are fleeing.

In Chapter 6, we ask why countries have had little success in cooperating on issues of migration. There are few regionally specific efforts to address migration, but there are international treaties that aim to impact state behavior. A large number of UN conventions provide guidelines for how various types of migrants—refugees, stateless people, economic migrants—should be treated. However, while refugees and migrant workers may be a global phenomenon, the costs and benefits of migration are much more localized than other transnational issues. Migrant workers from Indonesia and Rohingya refugees from Myanmar have few direct material impacts on these issues outside of humanitarian appeals. As a result, while UN conventions provide statements regarding the universal rights of migrants at the global level, most substantive efforts to address these issues occur at the regional level, whether multilateral or bilateral.

The book's conclusion, Chapter 7, summarizes our findings regarding the relative lack of effective cooperation on transnational security threats in Southeast Asia and offers some explanations on why this is the case. At a basic level, we argue that states find it difficult to cooperate when they are plagued by problems of poor governance domestically that make it difficult to carry out policies to address these problems, when interest groups are more concerned with their own private gain rather than the broader public good, and when regional or international organizations are too weak to promote effective cooperation from member states.

Notes

1. US Environmental Protection Agency, "Climate Change Basics," Washington, D.C., information snapshot as of January 19, 2017. https://19january2017snapshot.epa.gov/climatechange/climate-change-basic-information_.html. Under the Trump administration, information on climate change has been removed from the EPA website. The climate change information previously located on the EPA website can be found at the archived site listed above.

2. Ibid.

3. Zelda Soriano, "Climate Change Demands that ASEAN's Future Is Cooperative," *The Nation,* May 1, 2013.

4. Bernadinus Steni, "Climate Change and ASEAN," *Jakarta Post,* May 10, 2011.

5. World Health Organization, "Food Security and Famine Glossary of Terms," 2012, www.who.int/trade/glossary; Colin D. Butler, "Food Security in the Asia-Pacific: Climate Change, Phosphorus, Ozone and Other Environmental Challenges," *Asia Pacific Journal of Clinical Nutrition* 18, no. 4 (2009): 590–597.

6. S. Rajaratnam School of International Studies (RSIS) Centre for Non-Traditional Security (NTS) Studies, *International Conference on Asian Food Security 2011—Feeding Asia in the 21st Century: Building Urban-Rural Alliances (10–12 August 2011)* (Singapore: RSIS NTS, 2011), 7.

7. Organisation for Economic Co-operation and Development (OECD), "Share of ODA for Agriculture (percent)" (Paris: OECD Creditor Reporting System, 2009).

8. Asian Development Bank, "Asian Water Development Outlook: Measuring Water Security in Asia and the Pacific" (Los Baños, Philippines: Asian Development Bank, 2013).

9. National Academy of Sciences, *The Domestic and International Impacts of the 2009 H1N1 Influenza A Pandemic: Global Challenges, Global Solutions Workshop Summary* (Washington, DC: National Academies Press for Institute of Medicine, 2010).

10. International Organization for Migration (IOM), "Key Migration Terms," 2016, www.iom.int/key-migration-terms.

11. Alan Dupont, *East Asia Imperiled: Transnational Challenges to Security* (New York: Cambridge University Press, 2001).

12. This discussion builds on Terry Deibel, *Foreign Affairs Strategy: Logic for American Statecraft* (Cambridge: Cambridge University Press, 2014), 125–128.

13. Many analysts identify "value projection" as a fourth category of interests, but this set of interests is pertinent only largely to the rich and powerful, in contrast to the three other categories that are more universal. Moreover, the other three sets of interests can be challenged by traditional and nontraditional issues, which is not the case with value preservation and therefore it will not be discussed further.

14. There is a large body of literature on how states that want to get others to comply with their demands can issue effective threats, but that is outside the scope of this book.

15. Clearly, both threats and vulnerability are necessary to produce danger: vulnerability without a threat cannot create danger, but without vulnerability threats cannot impose costs, and therefore governments have few incentives to respond to it.

16. Deibel, *Foreign Affairs Strategy,* 149.

17. Shahar Hameiri and Lee Jones, *Governing Borderless Threats: Non-Traditional Security and the Politics of State Transformation* (London: Cambridge University Press, 2015); James H. Mittelman, *Hyperconflict: Globalization and Insecurity* (Stanford: Stanford University Press, 2010), 164.

18. Mutiah Alagappa, *Asian Security Practice* (Stanford: Stanford University Press, 1998), 624.

19. Arthur H. Westing, "The Environmental Component of Comprehensive Security," *Bulletin of Peace Proposals* 20, no. 2 (1989): 129–134, at 129.

20. Ralf Emmers, "Comprehensive Security and Resilience in Southeast Asia: ASEAN's Approach to Terrorism," *Pacific Review* 22, no. 2 (2009): 159–177.

21. Stephen Hoadley, "The Evolution of Security Thinking: An Overview," in *Asian Security Reassessed,* ed. Stephen Hoadley and Jurgen Ruland (Singapore: Institute of Southeast Asian Studies [ISEAS], 2006), 17.

22. Ibid., 18.

23. UN Development Programme (UNDP), *Human Development Report 1994* (New York: Oxford University Press for the UN Development Program, 1994), 23.

24. United Nations Development Programme, Human Development Report, 1994 (New York: Oxford University Press, 1994), p. 22. Quoted in Roland Paris, "Human Security: Paradigm Shift or Hot Air?" *International Security*, 26, no. 2 (2001): 89.

25. Mely Caballero-Anthony and Alistair D. B. Cook, eds., *Non-traditional Security in Asia: Issues, Challenges and Framework for Action* (Singapore: ISEAS, 2013), 1–14.

26. For a classic discussion of the public goods and ways to overcome the problems of collective actions, see Mancur Olson, *The Logic of Collective Action* (Cambridge: Harvard University Press, 1965).

27. For a good work on these concepts, see Angela Kallhoff, *Why Democracy Needs Public Goods* (New York: Lexington Books, 2011).

28. Helen Milner, "International Theories of Cooperation Among States," *World Politics* 44, no. 3 (1992): 468.

29. Robert O. Keohane, *After Hegemony* (Princeton: Princeton University Press, 2004).

30. See the different definitions and approaches in Friederich Kraochiwil and Edward D. Mansfield, eds., *International Organization and Global Governance* (New York: Pearson, 2006).

31. Michael Davies and Richard Woodward, *International Organizations* (Northampton, MA: Edward Elgar, 2014), 13.

32. Charles P. Kindleberger, "An Explanation of the 1929 Depression," in *The World in Depression, 1929–1939* (Berkeley: University of California Press, 1973), 291–308; W. Max Corden, "American Decline and the End of Hegemony," *SAIS Review* 10, no. 2 (1990): 13–26; Michael H. Hunt, "American Decline and the Great Debate: A Historical Perspective," *SAIS Review* 10, no. 2 (1990): 27–40; Helen Milner, "International Political Economy: Beyond Hegemonic Stability," *Foreign Policy,* no. 110 (1998): 112–123.

33. Kindleberger, "An Explanation of the 1929 Depression."

34. Kindleberger, "An Explanation of the 1929 Depression"; Corden, "American Decline and the End of Hegemony"; Hunt, "American Decline and the Great Debate."

35. Robert S. Chase, Emily B. Hill, and Paul Kennedy, eds., *The Pivotal States: A New Framework for U.S. Policy in the Developing World* (New York: Norton, 1999).

36. Commission on Global Governance, *Our Global Neighborhood: Report of the Commission on Global Governance* (New York: Oxford University Press, 1995), http://www.gdrc.org/u-gov/global-neighbourhood/chap1.htm.

37. Emmanuel Adler and Steven Bernstein, "Knowledge in Power: The Epistemic Construction of Global Governance," in *Power in Global Governance,* ed. Michael N. Barnett and Raymond Duvall (New York: Cambridge University Press, 2005), 294–318, at 302; and see James N. Rosenau, "Governance, Order, and Change in World Politics," in *Governance Without Government: Order and Change in World Politics,* ed. James N. Rosenau and Ernst-Otto Czempiel (Cambridge: Cambridge University Press, 1992), 1–29.

38. Margaret P. Karns and Karen A. Mingst, *International Organizations: The Politics and Processes of Global Governance,* 2nd ed. (Boulder: Lynne Rienner, 2009), 24.

39. Marco Schaferhoft, Sabine Campe, and Christopher Kaan, "Transnational Public-Private Partnerships in International Relations: Making Sense of Concepts, Research Frameworks, and Results," *International Studies Review* 11, no. 3 (2009): 451–474, at 451.

40. Michael Barnett and Martha Finnemore, "The Power of Liberal International Organizations," in *Power in Global Governance,* ed. Michael Barnett and Raymond Duvall (New York: Cambridge University Press, 2005), 162.

41. Katerina Tolarova, "Global Governance: Fragmentation, Stagnation, Transformation? A Talk with Miles Kahler," University College London, December 13, 2016.

42. Jessica R. Daniel, Sandra Lee Pinel, and Jase Brooks, "Overcoming Barriers to Collaborate Transboundary Water Governance: Identifying Local Strategies in a Fragmented Governance Setting in the United States," *Mountain Research and Development* 33, no. 3 (2013): 215–224.

43. Amitav Acharya, "The Future of Global Governance: Fragmentation May Be Inevitable and Creative," *Global Governance* 22, no. 4 (2016): 453–460.

44. Tolarova, "Global Governance."

45. Frank Biermann, Philipp Pattberg, Harro von Asselt, and Fariborz Zelli, "The Fragmentation of Global Governance Architectures: A Framework for Analysis," *Global Environmental Politics* 9, no. 4 (2009): 19–20.

46. Neda A. Zawahri and Sara McLaughlin Mitchell, "Fragmented Governance of International Rivers: Negotiating Bilateral Versus Multilateral Treaties," *International Studies Quarterly* 55, no. 3 (2011): 835–858.

47. Ibid., 835–836.

48. Ibid., 837. Faisal Hossain and Nitin Katiyar, "Improving Flooding Forecasting in International River Basins," *EOS, Earth and Space Science News, Transactions, American Geophysical Union* 87, no. 5 (2006): 49–60.

49. One of the best examples of these dynamics can be seen in the change in US policy toward the use of chlorofluorocarbons (CFCs) and the global environmental issue of ozone depletion. In the 1980s, there were only a handful of large producers of CFC-based products. US manufacturer DuPont was one of the largest of such producers. Once DuPont developed a cheap substitute for CFCs for many of its applications, it actively lobbied the United States to regulate, and ultimately heavily restrict, the use of CFCs domestically. This paved the way for US leadership in negotiations leading up to the Montreal protocol on the ozone layer in 1987. The Montreal protocol, which set targets for phasing out CFCs, halons, and other ozone-depleting substances, is widely held up as a model of international cooperation on an environmental problem. While US leadership was not the only positive ingredient that led to the signing and success of this sort of international cooperation, there is no question that it was a necessary element of it. Domestic political dynamics (the significant role of interest groups in shaping US policies) contributed to strong US leadership in the drafting, ratifying, and complying with the international agreement that would significantly reduce global use of ozone-destroying substances. This example shows that cooperation on issues of

human security is possible as long as there is leadership and there are incentives for cooperation. Marvin Sooros, *The Endangered Atmosphere: Preserving a Global Commons* (Columbia: University of South Carolina Press, 1997), 160–161.

50. David C. Dawe, "The Philippines Imports Rice Because It Is an Island Nation," in *Why Does the Philippines Import Rice? Meeting the Challenge of Trade Liberalization,* ed. David C. Dawe, Piedad F. Moya, and Cheryll B. Casiwan (Los Baños, Philippines: International Rice Research Institute and the Philippine Rice Research Institute, 2006).

51. Melody Aguiba, "Philippines to Play a Critical Role in Global Food Security with 2012 Golden Rice Launch," *Jakarta Globe,* March 19, 2011; Butler, "Food Security in the Asia-Pacific."

52. ASEAN Ministers on Agriculture and Forestry (AMAF) Plus Three, "Joint Press Statement of the Tenth Meeting of the ASEAN Ministers on Agriculture and Forestry and the Ministries of Agriculture of the People's Republic of China, Japan, and the Republic of Korea," Tenth AMAF Plus Three, Phnom Penh, October 24, 2010.

53. "Vietnam to Cooperate with Thailand on ASEAN Rice Cartel," *Angimex,* June 26, 2012, http://www.angimex-kitoku.com.vn/en-US/home.

54. Mekong River Commission, "Mekong River Commission History," 1995, www.mrcmekong.org; Mekong River Commission, "Agreement on Cooperation for Sustainable Development, Article 5," April 5, 1995.

55. Biermann, Pattberg, van Asselt, and Zelli, "The Fragmentation of Global Governance Architectures."

56. Ramona Alikiiteaga Gutiérrez, Monica Jane Naughtin, Srey Viseth Horm, Sorn San, and Philippe Buchy, "A (H5N1) Virus Evolution in South East Asia," *Viruses* 1, no. 3 (2009): 335–361.

57. Sophal Ear, "Avian Influenza: The Political Economy of Disease Control in Cambodia," *Politics and the Life Sciences* 30, no. 2 (Fall 2011): 2–19; Sophal Ear, "Emerging Infectious Disease Surveillance in Southeast Asia: Cambodia, Indonesia, and the US Naval Area Medical Research Unit 2," *Asian Security* 8, no. 2 (2012): 164–187; Elizabeth Wishnick, "Dilemmas of Securitization and Health Risk Management in the People's Republic of China: The Cases of SARS and Avian Influenza," *Health Policy and Planning* 25, no. 6 (2010): 454–466.

58. Gianna Gayle Amul, "ASEAN's Critical Infrastructure and Pandemic Preparedness," *NTSblog*, January 22, 2013, https://ntsblog.wordpress.com/2013/01/22/aseans-critical-infrastructure-and-pandemic-preparedness/.

59. United Nations High Commissioner for Refugees, "UNHCR Global Trends 2015," 2015, www.unhcr.org/en-us/statistics/unhcrstats/576408cd7/unhcr-global-trends-2015.html.

2

Climate Change: Complex Patterns of Conflict and Cooperation

Climate change is a global phenomenon that has diverse impacts on countries across the world and is hitting Southeast Asian countries particularly hard. As discussed in Chapter 1, *climate change* refers to any significant change in the measures of climate lasting for an extended period of time and includes major increases in temperature, precipitation, wind patterns, or other events that occur over several decades or longer. Climate change is a function of both naturally occurring and man-made factors, but in recent decades has been produced primarily by *global warming,* which refers to the ongoing rise in global temperatures caused by increasing concentrations of greenhouse gases in the atmosphere. The earth's average temperature has risen by 1.5 degrees F over the past century and is projected to rise another 0.5 to 8.6 degrees F over the next hundred years.[1] Global warming has produced changes in rainfall, droughts, more severe heat waves, melting ice caps, and transformations in oceans across the globe.

All Southeast Asian countries are vulnerable to the effects of climate change and should therefore share an interest in combating it collectively. Rising sea levels threaten to submerge cities such as Bangkok and Jakarta while the intrusion of saltwater into Vietnam's agricultural land is already reducing productivity and negatively impacting food security. Changes in rainfall have produced droughts, such as the horrific one in the Mekong region in 2015–2016, which threaten water and food security. Rising temperatures have produced new threats from invasive insects, like the caterpillar explosion in Java and Madura in 2011, or loss of cultivable land, which can lead to further food insecurity.[2]

Global warming, as discussed in Chapter 1, is a global public bad that negatively impacts all countries. Since it shares the characteristics of a global public good, there have been robust international efforts over the past two decades coordinated by the UN Framework Convention on Climate Change (UNFCCC) to overcome the collective action problem and promote cooperation among countries to reduce GHGs. At the December 2015 meeting of the Conference of Parties (COP21) in Paris, a historic agreement was reached that calls on all countries that sign the Paris Agreement to take actions to stem its GHG emissions as soon as possible with the collective goal of limiting the rise of global warming to 2.0 degrees C (3.6 degrees F) by 2100. The Paris Agreement entered into force on November 4, 2016, thirty days after at least fifty-five countries accounting for 55 percent of global emissions ratified the treaty.[3] Like many UN treaties, the Paris Agreement is a mix of legally binding and nonbinding elements; it requires countries to submit their national action plans for reducing GHGs, but lacks any mechanisms to compel countries to comply with their plans or sanction them if they fail to do so. In addition to efforts to mitigate the causes of climate change by reducing GHGs, the Paris Agreement addresses efforts to adapt to its effects. Since this book focuses on cooperative efforts toward transnational threats, in this chapter we concentrate on efforts to mitigate climate change rather than efforts to adapt to it that mainly occur at the domestic level, albeit often with external assistance.

In most countries GHG emissions are caused primarily by industry, but in Southeast Asia deforestation and changes in land use are key drivers of emissions. In Thailand, which began exploiting its forests earlier than other Southeast Asian countries, the percentage of land covered by forests declined from 53.33 percent in 1961 to 25.13 percent to 1998, a period during which the rate of deforestation exceeded 3 percent a year. As Table 2.1 illustrates, between 1995 and 2005, 43.6 million hectares were deforested in the region, which released about 3.45 million tons of carbon into the atmosphere according to the UN Food and Agriculture Organization.[4] Indonesia, Cambodia, Laos, Burma, Thailand, and Vietnam not only have lost old-growth forests but also mangrove cover, which has played a critical role in preventing soil erosion and therefore flooding in the past forty years.[5] In addition to its role in climate change, deforestation is contributing to loss of biodiversity, and it is posing problems for indigenous communities who rely on these ecosystems for their livelihoods.

Given the region's emission profile, in this chapter we focus on Indonesia's efforts to reduce its rates of deforestation and the related issue of preventing the transboundary haze caused by forest fires on peatland. We begin by examining Indonesia's role in the UNFCCC to illustrate that it has played a key mediatory role between developed and developing countries in climate change diplomacy and has made significant commitments under the UNFCCC. Indonesia's ability to meet its international commitments, however, has been hampered by a lack of technical capacity, poor governance in the forestry and land sectors, and efforts by power-

Table 2.1 Annual Forestation Levels by Country (in 1,000s of hectares)

	1990	2000	2005	2010	2015	Loss/Gain, 1990–2015 (%)
Indonesia[a]	116,567	97,852	88,495	94,432	91,010	–22
Vietnam[b]	9,363	11,725	12,931	13,797	14,773	+58[d]
Cambodia[b]	12,946	11,541	10,447	10,094	9,457	–27
Thailand	14,005	17,011	16,100	16,249	16,399	+17[d]
Laos[b]	17,314	17,532	16,142	15,751	18,761	–8
Malaysia[c]	22,376	21,591	20,456	20,456	22,195	–0.8
Philippines[b]	10,574	7,949	7,162	7,665	8,040	–24
Myanmar[a]	39,219	34,554	32,222	31,773	29,041	–26

Sources: FAO, "Global Forest Resources Assessment 2015" (Rome: FAO, 2015); FAO, "Southeast Asia Subregional Report, Asia-Pacific Forestry Sector Outlook Study" (Bangkok: FAO, 2011).

Notes: a. FAO, "Global Forest Resources Assessment 2010, Main Report," Forestry Paper No. 163 (Rome: FAO, 2010).

b. FAO, "Global Resources Assessment, 2005: Country Reports for Vietnam, Cambodia, and the Philippines" (Rome: FAO, 2010).

c. FAO, "Global Forest Resources Assessment 2010 Country Report: Malaysia" (Rome: FAO, 2010).

d. Although Vietnam has seen levels of overall forest cover increase, this increase has come in new forest cover, not in the primary forest cover, which has continued to be destroyed in Vietnam, particularly along the Mekong River. Similarly, the increase in the Thai forest cover comes from new planting.

ful private actors who benefit from the current system to thwart efforts to improve governance. We analyze these issues by focusing on the government's efforts to implement Reducing Emissions from Deforestation and Forest Degradation (REDD), restructure the chaotic and fragmented forestry governance structure, and enact a strong moratorium on forest conversion. We illustrate that despite significant international support and the magnitude of the threat that climate change poses to Indonesia's broad national interests, powerful private actors have largely been able to stymie effective action, as was made abundantly clear by the 2015 haze.

In this chapter, we therefore also examine efforts to halt the transboundary haze produced by forest fires in Indonesia that threaten the public health not only of its own citizens, but also those in neighboring countries, particularly Singapore and Malaysia. The haze is an extreme case of negative externalities, in which the actions of private actors in Indonesia impose tremendous health, social, and economic costs on neighboring countries. We examine the regional conventions created to address the haze, such as the ASEAN Agreement on Transboundary Haze Pollution, and discuss Singapore's efforts to hold Indonesian plantation owners accountable for their actions, which has triggered only conflict rather than cooperation among Southeast Asian states. We then discuss the new and potentially significant measures adopted by the Joko Widodo (aka Jokowi) administration in 2016 in response to the recent haze that, if implemented, could be potential game changers in Indonesian climate change efforts.

Indonesia as a Global Climate Actor

Indonesia's status as one of the world's largest producers of greenhouse gases as well as one of the countries most susceptible to its impacts makes Indonesia a pivotal actor in global efforts to combat climate change. Emissions vary year by year, but Indonesia consistently ranks as one of the world's top ten emitters, and some years has ranked as third-largest emitter of GHGs, trailing only the United States and China.[6] Indonesia's emissions profile, however, differs radically from the United States and China, where the majority of GHGs are produced by industry, transportation, and electricity. With the world's second-largest tropical rain forest and fastest deforestation rate, the bulk of Indonesia's GHGs come from deforestation and changes in land use.[7] Globally, only 20 percent of GHGs come from changes in land use, but in Indonesia that rate is 85 percent.[8] As a result, Indonesia's efforts to mitigate climate change are largely focused on REDD. Indonesia and Brazil together account for approximately 80 percent of global REDD potential, placing Indonesia on the cutting edge of global REDD efforts.

Indonesia is extremely vulnerable to climate change and has already begun to experience its negative effects. Experts predict that climate change will continue to produce dramatic impacts on agriculture, development, fisheries, and urban flooding as a result of changes in rainfall patterns, extreme weather events, and sea-level rise. Indonesia has already lost some of its 17,000 islands to rising sea levels and another 2,000 are at risk. Some estimates predict that Jakarta, the capital city that sits at sea level, could be underwater by the end of the century. Saltwater intrusion could render much of its arable agricultural land unproductive and trigger large-scale migration in some of the world's most densely populated areas. Changes in rainfall and rising temperatures will negatively impact agriculture, and the World Bank predicts that climate change could threaten food security in Indonesia.[9] Indonesia sits on the ring of fire and has long been prone to natural disasters such as earthquakes and tsunamis, but rising temperatures will produce more extreme weather events that will likely increase the number and severity of natural disasters. Indonesian forests contain the world's greatest biodiversity, and climate change threatens many endangered animal and plant species.

Given the magnitude of the threat that climate change poses to Indonesia as well as the significant role that it can play in mitigating it, under President Susilo Bambang Yudhoyono (2004–2014) the country took a leadership role in global climate change negotiations. In December 2007, Indonesia hosted the thirteenth meeting of the UNFCCC in Bali, which was convened to negotiate a successor to the Kyoto Protocol. The key fault line in UNFCCC negotiations has been between developed and developing countries. Developing countries argue that since industrialized countries are historically responsible for most GHGs, they should bear the burden of mitigating climate change. In contrast, industrialized countries argue that since current emissions growth comes largely from developing countries, they also must commit to limit GHGs. Overcoming the collective action problem regarding who should pay the costs of limiting GHGs by forgoing profitable activities has long been a key obstacle to reaching global agreements on climate change.

In 2007 in Bali, Indonesia brokered a compromise that called on all countries to play a role in mitigating climate change based on the principle of common but differentiated responsibilities, which was enshrined in the Bali Road Map. The "common" component was the recognition that a healthy atmosphere is a global public good and all countries have an interest and a responsibility to protect it. The "differentiated" component was a recognition that industrial countries had contributed most to global warming and, therefore, had to lead the way in reducing their emissions and in funding efforts by developing countries to mitigate climate change and shift to low-carbon development trajectories. The Bali Road Map also endorsed REDD for the first time, a milestone for the UNFCCC that had previously focused on industrial emissions. In 2009, Indonesia became the first developing country to make a firm commitment to reduce climate change when Yudhoyono pledged a 26 percent reduction in GHGs from business as usual by 2020, and up to 41 percent if Indonesia received significant international aid. More recently in Paris, President Jokowi pledged to cut its emissions 29 percent from business as usual by 2030 or 41 percent if granted significant international aid, demonstrating strong continuity with his predecessor.[10] It is important to note, however, that since business as usual is predicated on significant expansion of development in the agricultural, plantation, mining, and industrial sectors, a 29 percent reduction from that level means that Indonesian GHGs would still soar, nearly doubling from their 2010 levels.[11]

Indonesia's commitments and its potential to make a significant contribution to limiting GHGs have drawn international attention and support. The most significant is its partnership with Norway which, like

Indonesia, has positioned itself at the forefront of global efforts to combat climate change. In May 2010, the two countries signed a Letter of Intent (LOI) stating that the goal of both countries' climate change policies is to "ensure that they contribute *beyond their fair share*" to limit the average rise in global temperatures to less than 2 percent above the preindustrial baseline.[12] Norway has pledged up to $1 billion in exchange for verifiable reductions in Indonesian GHGs, which will be achieved primarily through REDD. However, as the following analysis of the key deliverables under the LOI—a two-year moratorium on forest conversion and a comprehensive monitoring system—illustrates, Indonesia has been unable to make good on its international commitments. Indonesia faces significant problems of technical domestic capacity and political will in halting the actions of powerful private actors, as the haze or pollution triggered by forest fires demonstrates.

Indonesia's Carbon Profile and REDD+

Indonesia is home to the world's second-largest tropical rain forest, which covers approximately 130 million hectares.[13] The Center for International Forestry Research (CIFOR), an independent research institution, contends that Indonesia is losing 1.1 million hectares of rain forests a year. Key drivers of deforestation in Indonesia are land use, weak spatial planning, tenure issues, ineffective forest management, poor governance, and weak law enforcement. Illegal logging, which is estimated to cost Indonesia $4 billion a year, is also a factor.[14]

Indonesia's forests are cleared not only for logging and the production of paper and pulp products, but also to convert land for agricultural businesses, mining, and energy projects. Half of Indonesian GHG emissions are released not simply from the cutting down of trees, but from the clearing of peatland under the forest. Much of Indonesia's ground cover is peat, which can extend many meters underground. Peatlands emit huge quantities of GHGs if drained for cultivation, which is why peat was traditionally used as fuel in places such as Ireland. Since 1999 it has been illegal in Indonesia to clear peatland over 3 meters deep, although this law is often broken. Peatlands are critical to fighting global warming because their carbon density can be five to ten times as high as forests on mineral soil per unit of forest.[15] When fire is used to clear forests in areas rich in peat, the fires burn underground for long after the forest is gone. The burning of peatland has been largely responsible for the haze that blankets parts of Indonesia, Malaysia, and Singapore each year and was particularly severe in 2015.

Preventing cultivation on peatlands will be key to Indonesia's success in meeting its climate change goals. Peat soil is anaerobic and acidic and, unlike mineral soils, is not well suited for agricultural purposes. Peatlands make up 50 percent of Indonesia's GHGs, but contribute only 1 percent to GDP, so the trade-off between reducing emissions and economic growth is significantly lower on peatlands than in other sectors.[16] Peatlands have been cultivated largely by palm oil companies, which presumably could be incentivized to cultivate other areas. The prominent place that peat occupies in Indonesia's carbon profile provides a significant opportunity to protect and restore forest while generating revenue through REDD.[17]

The goal of REDD is to make the conservation of forests, and therefore the sequestration of carbon or "avoided deforestation," as profitable as the drivers of deforestation.[18] It seeks to do so by altering the incentives for deforestation and land use change to encourage sustainable forestry management that will reduce GHGs. At base, countries like Indonesia will be paid in exchange for conserving forests and forgoing the economic gains that come from exploiting them. In 2008, the UNFCCC added three more strategic objectives to the original REDD goals of reducing emissions from deforestation and degradation: reducing emissions through conservation, sustainable forest practices, and enhancement of forest carbon stocks. The addition of these objectives has transformed REDD into REDD+.[19]

It is estimated that up to $30 billion a year could be made available to developing countries through global REDD+ mechanisms, and Indonesia is one of the countries with the potential to benefit most.[20] A UN report issued in September 2011 estimated that the value of conserving Indonesia's rain forests could generate three times more income than clearing them for oil palm plantations.[21] With respect to peat, because it is rich in carbon it should require a much lower price per carbon than other types of forests to entice landholders or license holders to participate in REDD+ projects. Peatlands are therefore some of the most cost-effective lands to engage in REDD+ activities. According to Agus Purnomo, who served as President Yudhoyono's chief climate change adviser and a member of Indonesia's National Council on Climate Change, Indonesia has the potential to earn up to $15 billion per annum by 2030 by reducing emissions.[22] In sum, REDD+ has the potential to be a win-win solution to overcoming the collective action problem on global warming.

As a completely new concept, however, putting REDD+ into practice requires the establishment of a wholly new institutional framework and the resolution of numerous technical and political issues. Four key challenges involved in making REDD+ work are measuring carbon, creating modalities for making payments, ensuring accountability, and funding. First, a prerequisite for setting a value on the amount of carbon-bearing potential of a forest is measuring the amount of carbon stored in it. Technologies such as satellite imaging and computer modeling can measure carbon stocks, making a transparent mechanism for verifying carbon stocks, and emissions possible. Second, debates over who should be rewarded for reductions in GHGs are contentious. Should external funds go to national governments, local governments, forest communities, or forestry companies who forgo logging? Third, accountability involves creating ongoing surveillance mechanisms to ensure that carbon payments produce sustained conservation. What recourse would exist if forests were exploited after a country received REDD+ funds? Finally, how should funding work? Should donor countries create a fund to reward countries that reduce forest-related GHGs? Or should there be a market-based system for carbon trading?[23] Or should there be some combination of the two? These contentious issues continue to be debated in global climate change negotiations.

At the same time, Indonesia, along with Brazil, the Democratic Republic of Congo, and other major forest countries, are engaged in a series of REDD+ demonstration projects that aim to prove that REDD+ (which is so logical in theory) works in practice. Proponents of REDD+ need to demonstrate that REDD+ is functionally viable to a number of different constituencies, illustrating the need for vertical cooperation among a diverse set of actors. First, sovereign donors who spend taxpayer monies must be able to prove that their aid is indeed paying for verifiable and lasting reductions in emissions, and not being lost to corruption or allocated for other purposes. This is particularly important for Indonesia, which is consistently ranked as one of the most corrupt countries in Asia. To reassure the international donor community, Indonesia has to demonstrate its capacity to establish transparent and accountable processes to monitor and verify emissions reductions.

At home, REDD+ proponents need to demonstrate not only the technical and administrative viability of REDD+, but also its long-term economic benefits. Indonesia, the world's fourth most populous country, is home to 250 million people, and ensuring food and energy security are important national interests that can conflict with climate change mitigation. Creating jobs for the 6 million new entrants to the labor force each year and raising the standard of living for the 50 percent of its population who live on less than $2 a day are also key social interests. By changing land use patterns, improving land use management, and shifting toward a low-carbon development plan, the government hopes to reconcile its goal of robust economic growth with a 26 percent reduction in GHGs.[24] Achieving these twin objectives will likely require difficult trade-offs.

There are many REDD+ skeptics in Indonesia, ranging from government officials, business leaders, and local forest communities being asked to forgo short-term economic gains in the hope of long-term economic, social, and environmental benefits. Many of these, particularly at the elite level, are corrupt officials and businesspeople who have personal economic interests in continuing business as usual. In addition, despite scientific evidence to the contrary, some people throughout the world continue to argue that global warming and climate change simply do not exist. In Indonesia, where many view international issues through the prism of a staunch nationalism and are suspicious of foreign ideas, climate change has sometimes been portrayed as a Western conspiracy designed to keep Indonesia down by preventing it from developing. Advocates of Indonesia's $16 billion a year palm oil industry have made this argument. For example, Achmad Mangga Barani claims to not believe in climate change, contending that "foreign countries saw palm oil as having the potential to trigger climate change, but so far there hasn't been accurate research that proves that."[25] Such sentiments could be dismissed if they were not spoken by the man who until recently was the director general for estate crops at the Ministry of Agriculture. Illustrating the tight links between government and business officials, on leaving his government position he became chairman of the Forum for Development of Sustainable Strategic Plantations, an industry lobby group. The ability of a powerful private actor to marshal political opposition to climate change mitigation efforts by mobilizing anti-Western sentiment complicates the task of promoting good governance.

At the local level, a lack of knowledge about climate change and carbon is an obstacle to REDD+ projects. According to former chairman of Indonesia's REDD+ Task Force, Kuntoro Mangkusubroto, forest communities simply do not understand climate change and convincing them to support conservation efforts is nothing less than a paradigm shift. He explains, "The paradigm in the past was you cut trees to get money, now we can change that paradigm and we can keep the trees and get money as well, but don't talk about carbon or they will not understand. Keeping water in the ground they understand, but what is carbon?"[26] Funds from REDD+ could play a pivotal role in alleviating poverty in rural areas, but according to a senior WWF official, "The basic obstacle is the community understanding of the issue."[27] Environmentalists have called for community outreach programs to raise awareness of REDD+ at the village level. In short, ensuring that all major Indonesian stakeholders support REDD+ is critically important, but extremely difficult given the fragmented nature of forestry governance.

REDD+ pilot projects need to overcome the concerns of skeptics at the international and domestic levels, but key steps necessary to reassure one constituency can upset the other. For example, the implementation of transparent accountable processes that are so important to international actors can be painstakingly slow. This delays the disbursement of funds to Indonesian actors who are taking steps to reduce GHGs and need to see the income potential of REDD+ in the short-term to offset their lost income.[28] The danger is that too long a period between the implementation of conservation efforts and compensation for local stakeholders will undermine conservationists in their battles against those who want to exploit the forests. In the absence of monetary compensation, the incentives for local actors to forgo exploitation decline. An increase in forest exploitation in turn, only weakens the incentives for international supporters to fund REDD+ efforts, triggering a potentially vicious circle that could undermine the viability of REDD+ at home and abroad. It was precisely to forestall such a scenario and demonstrate that REDD+ could work that Indonesia and Norway entered into a partnership in 2010.

Indonesia's Climate Change Partnership with Norway: Bilateral Cooperation to Support Global Commitments

In May 2010, Norway signed an LOI pledging up to $1 billion to Indonesia for verifiable reductions in GHGs from deforestation, land degradation, and peatland conversion that would be achieved largely by Indonesian's adoption and implementation of its REDD+ strategy. The partnership created by the LOI envisioned a three-step process, with the bulk of the funding coming in Phase 3 as compensation for verifiable reductions in GHGs.[29] In Phase 1, the preparation phase, Indonesia was to take steps necessary for the implementation of REDD. Specific deliverables in this phase include: the completion of a national REDD+ strategy that addressed all key drivers of forest and peatland related emissions; the establishment of a special REDD+ agency reporting directly to the president to coordinate efforts pertaining to the development and implementation of REDD+; the development of an independent institution for a national monitoring, reporting, and verification (MRV) system for forest and peat GHGs; and the design of a transparent funding instrument, managed by an internationally reputable agency, that would ultimately be the mechanism through which Indonesia receives funds for reducing GHGs.[30]

In Phase 2, the transformation phase, scheduled to begin in January 2011 and conclude by December 2013, Indonesia was to begin implementing the mechanisms designed in Phase 1.[31] Furthermore, Indonesia was to place a two-year moratorium on the conversion of peat and natural forest, establish a degraded land database, enforce existing laws against illegal logging and trade in timber, take measures to address land tenure conflicts, and create a province-wide pilot for REDD+.[32] It was envisioned that Phase 3 would begin in 2014, at which time Indonesia would begin receiving funding.

Progress on Indonesia's end has been rocky. Technical and political difficulties have plagued Indonesia's efforts to build its forest governance infrastructure and reduce GHGs. In fact, since the partnership was signed, satellite imagery has shown that Indonesia's deforestation rate continued to increase. Since no progress was made on decreasing deforestation, no performance-based payments were made. Norway released funds for preparation phases and for the transformation phase, accounting for less than $50 million.[33] The Norway LOI remains in place and still provides an incentive for efforts to reduce deforestation,

The two-year moratorium on land conversion, the most important tool in the efforts to limit deforestation, is also the most politically contentious issue in the agreement. The announcement of the moratorium triggered reactions from a wide range of Indonesian government actors, business interests, and societal groups, precisely because it had the potential to impact a wide range of interests. Broadly, business groups (particularly in the agriculture, mining, and energy sectors) opposed the moratorium, arguing that it would hamper economic growth. Environmental groups, on the other hand, largely welcomed the deal, as did many advocates for indigenous peoples who depend on the forests for their livelihoods. In the next section, we discuss key aspects of forest governance in Indonesia. We analyze the technical challenges facing Indonesia's efforts to rationalize its forestry governance, the overlapping governmental authority that needs to be coordinated, and the political jockeying between different governmental ministries and private actors over the moratorium and the development of its spatial forest mapping system. We trace the evolution of Indonesia's efforts to build the technical and regulatory capacity to make REDD+ work, the gross failures that are evident in the 2015 peat fires, and the significant policy shifts taken by the Jokowi administration in late 2015 and 2016 that, if implemented, hold the potential to help Indonesia meet the climate change challenge.

Public Ownership, Private Use, and Classifications of Indonesia's Forests

In Indonesia, forests and other natural resources are publicly owned and according to law should be governed for the benefit of the people. By nationalizing all natural resources, Indonesia's founding fathers sought to ensure that the country's natural wealth be treated like a public good that should benefit the broad national interest, not private ones as under the colonial regime. Since all land is public (not private) property, the government has a powerful tool when granting concessions to private actors to use the forest. In theory, the government could condition concessions and the attendant right to profit from the forest on sustainable practices which, if not followed, could result in the revocation of that forest concession. In practice, however, Indonesian authorities have often granted concessions to politically connected businesspeople in exchange for political favors or economic kickbacks.

The announcement of a two-year moratorium on forestry conversion triggered a debate over which types of forests would be covered by the moratorium. The LOI used the term "natural forest," which the FAO defines as "a forest composed of indigenous trees and not classified as forest plantation."[34] The wide scope of this term, which differs from Indonesia's official classification system as well as the language often used in many NGO and press reports, created an opportunity for lobbyists on both sides of the issue to attempt to influence the shape of the moratorium. Since much of the political contestation focused on which different types of forests would be covered in the moratorium, in this section we briefly review different types of forests and Indonesia's classification system. It must be noted, however, that in practice the boundaries between these different types of forests are often unclear.

The Indonesian government has traditionally classified forests into three different categories—production, conversion, and conservation—with different activities permitted in each category. *Production forests* are areas designated for sustainable forest management and expected to maintain forest ecosystems within the forest estate. Production forests can be converted to other functions such as agricultural estate crops such as palm oil, sugar, and coffee. The decision to release land from the forest estate for other purposes is made by the Ministry of Forestry in response to a request from industry or a government agency. Once released, this land becomes *conversion forests*. Converted land is subject to land use decisions that are largely the authority of local governments or the National Land Agency.[35] *Conservation forests* are those not designated for exploitation and protected by law, although enforcement of conservation laws is extremely weak.

Other terms are often used to describe the extent to which forests have been subject to human activities. The FAO defines "primary forest" as naturally generated forests of native tree species, in which there are no visible signs of human activity and ecological processes are not significantly disturbed.[36] Indonesia uses this term in its reporting to the FAO to mean forests characterized by the absence of logging roads, regardless of whether the area is covered by a concession license. "Secondary forests" are defined as forests that have been cleared of much of their original vegetation and where trees are regenerating. Secondary forests make up more than half of Indonesia's forest area and are home to a large number of endangered species.[37]

Finally, once a forest has been open to at least one round of logging, it is considered to be a *degraded forest* and becomes convertible to agricultural crops such as oil palms. Precisely how much human impact and environmental damage needs to be done before land can be designated degraded is a matter of dispute and the lack of a common operational definition plays into the hands of businesses. Environmentalists claim that much land designated as degraded is only slightly altered by logging and contains significant carbon stocks and biodiversity. According to Greenpeace spokesman Ian Duff, "A weak and unspecific definition of what 'degraded' means plays into the hands of companies that want to continue business as usual, expanding into carbon-rich forests and peatlands."[38]

An integral part of REDD+ is to shift commercial activities away from primary forests to degraded land, and a key deliverable of the Norway LOI is to compile a database of degraded land to encourage development on it. But if much degraded land is carbon-rich, then encouraging plantations on it would mean clearing carbon-rich forests for plantations, precisely the opposite goal of REDD+. The World Resources Institute is attempting to devise criteria for identifying degraded land and have proposed four indicators for it: whether the land is environmentally degraded, economically viable, socially desirable, and legally available.[39] Although extended discussions of forestry classifications, criteria for identifying degraded land and databases to map peatland may make for dry reading, these technical issues are fundamentally important to Indonesia's efforts to reduce GHGs. As Indonesia moves forward with REDD+, the criteria it uses to classify forests will significantly impact whether shifting economic activities to it will generate the hoped-for GHG reductions.

Weak Forestry Governance: Overlapping Authority, Regulatory Uncertainty, and Corruption

Indonesia's forest concession allocation system has long been chaotic, which makes accounting for all licenses issued for forest use difficult. At the national level, the Ministry of Forestry, which Jokowi merged with the Ministry of Environment, allocates timber concessions to private industry. Once land has been converted to agricultural, energy, or mining purposes—the largest commercial activities conducted in forests—the land then comes under the regulatory authority of the Ministries of Agriculture or Mines and Energy. Competition between these ministries for access to forests can be rife and, in Indonesia's notoriously inefficient and corrupt bureaucracy, creates opportunities for private interests to influence policymakers.

Indonesia's Ministry of Forestry earns $15 billion a year from issuing forest permits to investors and has a long history of graft.[40] According to Chandra Hamzah, then deputy director of Indonesia's highly respected Corruption Eradication Commission, the Ministry of Forestry is "an unlimited source of corruption."[41] Indonesian Corruption Watch, an NGO, agrees, noting that illegal logging and violations of permit issuance guidelines produce an estimated $2.3 billion a year.[42] Doubts about Indonesia's capacity to implement REDD+ projects according to international standards of transparency and accountability were underscored by the naming of Wandojo Sisawnto, an Indonesian negotiator at the 2009 Copenhagen climate talks, as a bribery suspect and the jailing of forestry officials responsible for carbon offset projects for taking bribes.

Indonesian forest governance is characterized not only by vertical competition between different ministries, but also by horizontal competition between the central and local governments. Decentralization has granted Indonesia's 33 provinces and over 500 districts significant authority to grant licenses and otherwise affect land use. Oil palm plantation and mining permits, for example, were often granted by local authorities. District governance varies widely in terms of technical capacity, commitment to good governance, and access to credible information required for sound policymaking. Local authorities are often under pressure to raise revenue to fund local services or may be subject to pressure from local communities or businesses to accelerate forest exploitation. And some may simply be out for personal gain. Unclear or conflicting policies, discretionary decisionmaking, and opportunities for enrichment through corruption have led to the conversion of protected forests, overlogging, establishment of plantations outside of permitted areas, and other practices that have brought about uncontrolled deforestation.[43]

Moreover, the overlapping authority between different bureaucracies and levels of government meant that concessions to the same land were often granted by different government authorities to different private actors. As a result, attempting to identify which company was responsible for environmental devastation is difficult. As we discuss below, a key obstacle to combating the transboundary haze is determining which companies hold licenses to specific concessions where fires occur, and therefore can be held accountable. To address this problem, Indonesia is developing its One Map, which is discussed below.

Political Will and the Challenge of Intergovernmental Coordination

Presidents Yudhoyono and Jokowi have each demonstrated a commitment to redressing the coordination problems that complicate forestry governance, but have gone about it in different ways. President Yudhoyono viewed the issue as part of his political legacy and put his personal prestige on the line for climate change, declaring that he would "dedicate the last three years of my time as President to deliver enduring results that will sustain and enhance the environment and forests of Indonesia."[44] He made this pledge at a major conference hosted by CIFOR, a nonprofit scientific organization, indicating that he clearly supported the efforts of CIFOR and was open to international technical and financial assistance, in contrast to other Indonesian actors who have been more suspicious of foreign businesses and NGOs.

Even with strong support from the top, implementing climate change policy in Indonesia faces a number of serious obstacles ranging from intergovernmental coordination, bureaucratic obstacles, center-periphery issues, and the pecuniary interests of individual bureaucrats. Yudhoyono was fully aware of these obstacles as well as the fact that international support for his efforts would not be forthcoming if they could not be overcome.[45] Yudhoyono's solution was to create new institutions under the office of the president. Yudhoyono had created the National Council on Climate Change in 2008 under the executive branch and staffed it mainly with a group of environmentalists, technical advisers, and other reformers to advise him on the issue. A key task of the National Council on Climate Change was to craft Indonesia's position on international climate change negotiations. The Norway partnership was negotiated largely by the president's climate change advisers.

Under the Norway deal, Indonesia committed to establish a REDD+ Agency. It was finally established in 2014, and housed within the Presidential Work Unit for Development Monitoring and Control (UPK4). Indonesia's REDD+ Agency was the world's first cabinet-level agency devoted to climate change. UPK4 had been created earlier by Yudhoyono as a coordination tool for the president to overcome bureaucratic logjams and infighting that hampered the delivery of sound policy across the government writ large. It was headed by Kuntoro Mangkusubroto, the highly respected official who had earlier served as head of the Aceh rehabilitation agency in the wake of the devastating 2004 tsunami, where he earned praise from the international community and domestic actors alike.

Yudhoyono's attempts to centralize control over coordination of climate change and REDD+ policies under the office of the president came under criticism. The National Council on Climate Change and the REDD+ Agency had no independent authority over other ministries. As a result, they could identify problems and make recommendations, but lacked the authority to implement them. Many argued that the proliferation of presidential committees and task forces under Yudhoyono only complicated the coordination problem further.[46] Ministries complained that they were often forced to waste valuable time responding to questions and requests for data from multiple sources.

Jokowi took a different approach. He merged the Ministries of Forestry and Environment to improve management of Indonesia's forests, and in June 2015 Siti Nurbaya Bakar was appointed head of the agency. Moreover, Jokowi dissolved the National Council on Climate Change, although some of its members were appointed as advisers to the new ministry. Similarly, the REDD+ Agency was folded into the new ministry. The argument was that, by centralizing the responsibilities in one agency, policy coordination would be improved and bureaucratic overlap would be reduced. It is too early to tell if this streamlining will help or hurt environmental goals and programs.[47]

Reaction to the Norway Partnership and Moratorium Politics: Key Actors and Interests

The reaction to the Norway LOI and its proposed moratorium was swift from bureaucratic, business, and civil society groups. Climate change policy impinges on many diverse economic sectors, resulting in competing priorities between different ministries, the public and private sectors, and different levels of government. The debates over the moratorium illustrate the various interests, relative power, and coordination problems in Indonesia. Here, we discuss the two-year 2011 moratorium, which was renewed in 2013, 2015, and 2017. The subsequent regulations on forestry use issued by the Jokowi administration following the 2015 haze are discussed below.

The partnership was negotiated by the National Council on Climate Change, and some of the partnership's provisions, particularly the two-year moratorium on forestry conversion, caught other ministries by surprise. This was particularly true of the Ministry of Agriculture. Agriculture accounts for 45 percent of GDP and 45 percent of employment in Indonesia, and changes in land conversion could therefore have large social and economic impacts. The ministry wanted clarification on how the moratorium would affect its plans for sugar self-sufficiency, among other issues. The agricultural sector is far from monolithic. Small-scale farmers and indigenous peoples who rely on local forests for their livelihoods have very different interests from large-scale plantation agriculture, especially palm oil producers. The tremendous expansion of the palm oil sector in recent years has often pitted local communities dependent on the use of traditional forests, but who lack legal title to it, against powerful agribusiness interests. Resolving land tenure issues to ensure that REDD+ does not disenfranchise such groups is a key interest of many Indonesian political reformers, NGOs, and international donors.

Indonesia is the world's largest producer of palm oil, an industry that has been expanding rapidly in response to soaring demand from China and India. Indonesia's trade minister said that palm oil exports grew 16 percent in 2011 amid increased expansion in the industry; by 2014 and 2015 Indonesia was producing 31 million tons and exporting about 21 million tons of that. Production and exports have dropped since then due to the global and Chinese economic slowdown and aging trees in Indonesia.[48] The industry employs 3 million people, and advocates for it argued that the moratorium would harm economic growth and threaten jobs. Plantation owners prefer to clear virgin forests for plantations because it is cheaper and they reap profits from selling the timber. The Virginia-based World Growth Group, funded largely by a group of Malaysian palm oil producers, claimed that Indonesia would lose 3.5 million jobs every year if the country put any limitations whatsoever on forestry exploitation.[49] The Indonesian Palm Oil Association sent a delegation to Oslo to lobby against the moratorium.[50]

Proponents of the moratorium disputed that its climate change policies would slow economic growth by pointing to Brazil, which at that point had slowed deforestation by 70 percent while maintaining high rates of economic growth. Under Yudhoyono, the climate coalition's slogan was "7/26" in reference to its claims that Indonesia can grow at rates of up to 7 percent per year and still reduce GHGs by 26 percent. It adopted a mantra that its climate change policies were "pro-growth, pro-jobs, pro-poor, and pro-environment."

Environmentalists argue that cleaning new land to expand palm oil production was unnecessary since plantations could increase productivity from existing cleared areas. Average palm oil productivity in Indonesia is 3.5 tons per hectare, which is 40 percent less than Malaysia's average of 6.4 tons per hectare. Increases in productivity could negate the need to expand plantations and limit deforestation and the emissions that accompany it. Jokowi has established (with support from UNDP) a National Palm Oil Platform with the express goal of raising productivity, particularly for small producers. The government has created the Crude Palm Oil Supporting Fund (CPO Fund) to promote and facilitate this.[51]

The environmental lobby in Indonesia has largely welcomed the Norway partnership. Some have argued that it did not go far enough, but praised it as an initial first step. There have been exceptions. Displaying the third world solidarity that runs deep in Indonesia, the Indonesian Civil Society Forum for Climate Justice argued that the moratorium was premature because, according to the Kyoto Protocol, only developed countries are responsible for reducing emissions. One of Indonesia's oldest environmental groups, the Indonesian Forum for the Environment (WAHLI), opposes REDD+. It argues that REDD+ focuses too much on mitigation and not enough on helping local communities adapt to climate change. It also fears that REDD+ projects will undermine the rights of indigenous and forest communities, who often lack title to land that has been used as a source of their livelihoods for generations.[52]

Moratorium Politics

The political battle over the moratorium demonstrates the intense competition over deforestation in Indonesia. Under the LOI with Norway, Indonesia had agreed to establish a REDD+ Task Force to draft the moratorium and it was created in 2011. It was headed by Kuntoro, who was then head of the UPK4, under which the REDD+ Task Force was housed. With the moratorium scheduled to take effect only six months after the signing of the Norway deal, lobbyists on all sides began campaigning to ensure that their interests were reflected in it. Powerful business groups, particularly Indonesia's $16 billion a year palm oil industry, opposed the moratorium.[53] The political contestation became so intense that the moratorium was ultimately delayed for six months.

Given Kuntoro's background as a reformer, he was viewed as a natural ally of those seeking better forest governance. He sought input from a wide range of stakeholders, consulting with civil society groups, investors, and scientists among others. Some commended this as an attempt to ensure buy in from key stakeholders, but others, including a spokesman for the Nature Conservancy, argued that "it was a case of too many cooks spoiling the broth."[54]

Key points of debate were: What types of forests should be covered by the moratorium? Should it apply to existing permits or only to new ones? And if it were to apply retroactively, would license holders be compensated and, if so, how? Compensating license holders raised the unpalatable prospect that public funds could be used to compensate holders of permits obtained through corruption.

There was political conflict within the executive branch over the moratorium. The Forestry Ministry proposed that the moratorium only ban new permits in primary forests for two years while Kuntoro and many allies in the environmental movement wanted to extend the moratorium to secondary forests, review existing permits, and consider extending its time frame.[55] The draft moratorium produced by Kuntoro and the REDD+ Task Force called for a review of existing permits, and only extended license holders who had begun work under their permits by December 2010.[56]

In reaction to the moratorium produced by the REDD+ Task Force, the business community mobilized against it and began lobbying Forestry Minister Zulfiki Hasan, Environment Minister Gusti Muhammad, and Economic Coordinating Minister Hatta Rajasa. These officials produced an alternative draft moratorium. Hasan argued that the moratorium should only apply to new permits to maintain legal and regulatory certainty. According to Hasan, "Businesses who already hold permits can go ahead, we cannot and may not stop them because we have exploited (the forest) for 40 years, so there are many companies who have permits."[57]

On January 1, 2011, the date the moratorium was scheduled to come into effect, Yudhoyono had two competing moratorium drafts on his desk. As the deadlock continued, Yudhoyono tasked Vice President Boediono with reconciling the two drafts. Uncertain of the content of the moratorium and fearful of its impact, many businesses, in cooperation with their political and bureaucratic allies, used the delay in the moratorium to secure new licenses. In the last week of December 2010, when it was expected that the moratorium would take effect January 1, 2011, somewhere between thirty and forty licenses were issued "in principle," compared to a normal rate of three or four a month.[58] In the period between the moratorium's targeted start date, January 1, 2011, and its issuance in late May, the minister of forestry issued approximately forty additional new permits.[59]

The Moratorium Goes Into Effect

On May 20, 2011, just a week before the first anniversary of the LOI, the government issued Presidential Instruction (Inpres) No. 10/2011, through which the moratorium came into effect. An Inpres is a set of presidential instructions to specific ministries and government agencies. Since the Inpres is not a law, there are no legal consequences if the instructions are not implemented. In Indonesia, many government policies are enacted through Presidential Instructions rather than laws made by parliament because high political contestation and corruption in parliament makes it difficult to get sound laws passed. Jokowi used the same mechanism to extend the moratorium in 2015 and 2017.

This Inpres issued instructions to the Ministries of Forestry, Home Affairs, and Environment as well as five agencies: the Presidential Delivery Unit, the National Coordination Agency for Spatial Planning, the National Coordination Agency for Survey and Mapping, and the proposed REDD+ Agency.[60] The Inpres also applied to provincial governors and district heads, who had gained the authority to issue forestry licenses as a result of decentralization a decade earlier. Reflecting Indonesia's concern with food and energy security, neither the Ministry of Agriculture nor the Ministry of Energy and Mining were included in the Inpres, despite the role that agriculture and mining activities play in deforestation and GHG emissions. These exemptions meant that the implementation of the moratorium had to be made compatible with another Presidential Instruction, Inpres No. 5/2011 on food security, which instructs the minister of agriculture to expand the area of food production, the minister of forestry to allow the use of forestland for that purpose, and the minister of public works to build infrastructure to support rice production and transportation.[61] It is unclear by what process and under whose authority competing demands for forestry use to meet Indonesia's climate change, energy security, and food security goals will be resolved.

The Inpres applies to "primary natural forests" and peatlands, not "natural forests," the term used in the LOI.[62] As discussed below, the use of this new term generated criticism by environmentalists and controversy over precisely how much forest was covered by the moratorium. The Inpres did, however, take a strong stand on peatland, covering all peatlands regardless of their type, depth, location, jurisdiction, or level of disturbance.[63] Previously, only peatlands deeper than 3 meters were protected from conversion. Given the large percentage of GHGs caused by clearing peatland, this provision of the moratorium has the potential to significantly reduce emissions if implemented effectively.

The Inpres, however, included a number of exemptions that reduced its impact. First, it exempts forests covered by applications for concession licenses already approved in principle by the minister of forestry, regardless of their carbon density or biodiversity. Second, it exempts land needed for "vital" national development projects, defined as geothermal, oil and natural gas, electricity, rice, and sugarcane. As CIFOR notes, activities necessary to meet the demand for more electricity could be interpreted to mean the expansion of coal mining.[64] Third, the Inpres permits the extension of existing licenses for forest use as long as the license for business remains valid, regardless of the license holder's performance under an existing license or its environmental integrity. Finally, the fourth exemption creates a new opportunity to enhance carbon stocks through afforestation and reforestation. Ecosystem concessions would grant rights to the land for sixty-five years, with a possible thirty-five-year extension, for projects designed to restore the forests, an arrangement that could help address the permanence and accountability issues required for REDD+.[65]

The Reaction to the Moratorium

The issuance of the moratorium after such a long delay and intense political jockeying triggered a wide range of responses and much debate over the area of forests covered by it. In the wake of the moratorium, administration officials went to great lengths to deflect criticism that the moratorium was dictated by Norway and, hence, represented an Indonesian submission to foreign pressure. One presidential adviser claimed, "The moratorium is not a result of the deal with Norway."[66]

Environmental groups criticized the moratorium for a number of reasons. First, they argued that a two-year moratorium matters little. Instead, it should be performance based and last as long as necessary. Second, they contended that the use of the term "primary natural forest" was left deliberately vague as a concession to business. These are problems that would surface again in the context of Indonesia's handling of the haze in 2015. The term "primary natural forest" reinforces the interpretation that the moratorium targets only untouched, unmanaged, and undisturbed forests. Greenpeace also contended that since the primary forests designated in the decree were located mostly in forests already protected under law, there was less incremental protection than expected.[67] Many environmental groups had wanted the moratorium to cover secondary forests, and Greenpeace argued that the moratorium should have covered more than twice as much forest as covered in the decree. CIFOR, while noting that the failure to extend the moratorium to secondary forests was a lost opportunity, nevertheless praised the decree as a positive development for protecting forests and reducing emissions. In particular, it praised the inclusion of peatlands, noting that "the main positive environmental impact of the moratorium stems from the additional protection" it extends to them.[68]

The exemptions from the moratorium for existing concession licenses that had been approved in principle before the Inpres was announced generated controversy. Given the large number of licenses granted by the forestry minister in the run-up to the moratorium and the lack of clarity in defining "in principle," many fear that this language has the potential to severely undercut the moratorium's aim. Activists also criticized exemptions that would permit the conversion of forests for geothermal projects, rice and sugar plantations, and ecosystem projects.[69] But these exemptions had been announced at the time of signing of the LOI and justified as supporting the government's commitment to renewable energy, hydroelectric infrastructure, and reservoirs, which are other legitimate national interests.

Bitter disagreements broke out over who was to blame for this frenzy of licenses. Many environmental activists blame greedy businessmen and venal bureaucrats. But that view is challenged in part by Agus Purnomo, who served as an adviser to President Yudhoyono on climate change. He argues that by proposing a moratorium that went so far beyond what was intended in the LOI, particularly the proposal that the moratorium cover secondary forests, the environmental lobby created perverse incentives for forestry businesses to make a mad rush for licenses. According to this view, by moving beyond the scope of the politically possible to seek a perfect, but unworkable moratorium, the environmental lobby undermined its own cause.[70]

Many businesses, particularly concession holders and oil palm growers, are concerned that the moratorium will slow growth and threaten jobs. According to the US Department of Agriculture (USDA), citing figures from the Indonesian Palm Oil Association, the moratorium could slow the expansion of the area under cultivation from its average growth of 350,000 per annum over the past four years to 200,000 per annum. Oil palm plantations currently occupy 2.82 million hectares of peatland, but the moratorium restricts access to an additional 3.7 million hectares allocated for nonforestry development that otherwise may have been used.[71]

Many analysts dispute this contention that jobs, expansion, or economic growth are at risk. In January 2011, the Forestry Ministry had identified 35 million hectares of degraded land that could be used selectively for business, and nine plantation firms had already submitted requests to use 320,000 hectares. The Indonesian Palm Oil Producers Association (Gapki), which led much of the business lobbying against the moratorium, welcomed this announcement. According to Gapki's executive director, the designation of 35 million hectares of land would provide for future growth of the industry since "our rate of expansion is only 400,000 per annum."[72] Moreover, many large palm oil companies have large land banks consisting of granted or approved-in-principle licenses that will permit them to expand well beyond the term of the moratorium. For example, Gozco Plantations has a land bank of 124,000 hectares of which only 35,000 hectares have been planted, while Wilmar International has more than 570,000 hectares of land of which only 235,000 hectares have been planted.[73] The land banks of these two companies alone are sufficient for the palm oil sector not only to expand at its traditional rate of 350,000 hectares this year, but also to meet its more ambitious growth rate of 400,000 hectares.

Mapping the Moratorium

The 2011 moratorium was a significant step in Indonesian efforts to grapple with deforestation and reduce GHGs. As the political jockeying illustrates, however, Indonesia has other national interests, including food and energy security that conflict with reducing GHGs. Coordinating the various national-level ministries, managing central-periphery relations, and holding private actors accountable for their actions are all critically important tasks of immense proportions in Indonesia, illustrating the difficulties of overcoming the complexities of fragmented governance. Any moratorium, after all, is only as effective as efforts to implement it, and weak efforts to crack down on illegal logging and corruption in the forestry sector have continued to plague the government.[74]

A necessary, but heretofore nonexistent, tool in Indonesia's quest for sustainable forestry governance is a map that accurately surveys its forests according to type and locates the concessions that have been granted for their use. The lack of such a map resulted in conflicting estimates of how much forest was covered by the moratorium. In January 2011, Forestry Minister Hasan claimed that the moratorium would protect 43.8 million hectares of primary forests, and apply sustainable development practices to another 48 million hectares.[75] However, the Presidential Instruction on the moratorium was accompanied by an indicative moratorium map (IMM), which was released for public review in August 2011 by the Forestry Ministry. The IMM revised the area of primary forests to 55 million hectares, of primary forest, down from the 64 million earlier announced by Purnomo, illustrating the critical need for good data collection and transparency regarding forest use for overcoming the collective action problems associated with reducing emissions from deforestation.

To remedy this problem, Indonesia launched its One Map initiative, which aimed to create a single all-encompassing map that would contain all information relevant to forest licensing and land use claims.[76] The One Map attempts to reconcile the competing land use maps issued by different government authorities that show overlapping concession areas granted to multiple actors for different purposes. It seeks to gather the maps used by the Ministries of Forestry, Land Use, and Agriculture as well as those used by local governments and harmonize them. The One Map was first discussed in a 2010 when Yudhoyono and his advisers were presented with widely disparate maps from the Ministries of Forestry and Environment and realized that a single accurate map to track forestry use would be necessary to make REDD+ work. The One Map also serves the much broader national objectives of reconciling land tenure conflicts and providing better information for infrastructure development, disaster management, and a host of other issues.

The One Map initiative builds on the IMM that has been updated every six months, and is coordinated by Indonesia's Geospatial Information Agency (BIG), which has the technical expertise to produce the map. The map, however, is intensely political. As one authority has observed, "One of the main obstacles to developing a single map is the reluctance of government agencies to collaborate, particularly those with the authority to issue permits. With the One Map, these agencies will still have the authority to issue permits, but their opportunities for gaining illicit benefits will be greatly reduced."[77] For years, corrupt officials in the production orientation sectors (e.g., agriculture, mining, and logging) as well as local officials have benefited from unclear, contradictory, and inconsistent regulations. The fact that different agencies use different maps with inconsistent and outdated data complicates law enforcement.

In December 2014, BIG issued a basic geospatial map, along with several thematic maps such as a mangrove map, a maritime characteristic map, and a seagrass map. In February 2016, President Jokowi also issued a Presidential Regulation (Perpres) to accelerate the completion of One Map by 2019.[78] Any map, however, is only as good as the data on which it is based, and the data on Indonesian land use comes from agencies that often have an incentive to withhold it. Furthermore, holding private companies responsible for deforestation requires not only accurate concession data from government agencies, but also accurate information on the activities conducted in those concessions by private companies. As the discussion of the horrific 2015 haze illustrates, however, private actors have an incentive to withhold that information so as not to be held accountable for fires and other impermissible activities, once again illustrating the divergent interests of key actors whose compliance with forestry regulations is necessary to reduce emissions.

Transboundary Haze: Negative Externalities, Weak Cooperation, and Unilateral Attempts to Impose Costs and Enforce Compliance

A defining feature of transnational security issues is that they cross borders to threaten neighboring countries. The transboundary haze produced by peatland fires in Indonesia during the dry season that threatens the health of millions in Indonesia, as well as in Malaysia and Singapore, is a textbook case of how the actions of actors in one country create negative externalities that impose significant costs on neighboring countries. Pollution from forest fires typically started illegally to clear land has the greatest impact on those living close to them, usually Indonesians. Depending on the site of the fires, which mostly occur in Kalimantan, Sumatra, and Riau, and the prevailing winds, the haze typically engulfs Singapore and parts of Malaysia, and sometimes southern Thailand and the Philippines rather than Jakarta. Singapore and Malaysia are impacted because they are downwind of the fires. Indeed, it is often argued that if the haze affected the political and economic elite in Jakarta, government officials might summon the political will to respond more effectively to it.

The 2015 haze was the worst in twenty years, causing immediate threats to the health, social, and economic security of people living in the region.[79] Estimates show that on some days daily emissions from the fires surpassed the average daily emissions of the entire US economy. In the immediately affected areas, the Pollution Standard Index reached 2,000. Since anything over 300 is considered dangerous to public health, the haze triggered evacuations of the immediately affected areas in Indonesia.[80] Over 500,000 people were treated for acute respiratory tract infections and haze-related illnesses.[81] Throughout the region, schools and businesses were closed and transport was halted. Approximately 5 million students were affected by school closures. The World Bank estimates that the haze caused $16 billion of damage, an amount equal to 1.8 percent of Indonesia's GDP.[82]

Singapore and Malaysia suffered many of the same effects. Tourism was hard-hit in both countries when other governments issued health warnings. One early estimate was that the economic costs could hit 0.4 percent of Singapore's GDP.[83] The worst haze prior to that of 2015 occurred in 1997 when the region also was affected by El Niño, which delayed the monsoon rains that help put out the fires. In 1997, the haze cost the region an estimated $9 billion and there were fears that the 2015 haze would surpass that. In 2016, ASEAN authorized a study of the impact of the 2015 haze.

The magnitude of the threat posed by the 2015 haze triggered strong domestic and international criticism of Indonesia. Indonesian environmental groups and ASEAN ministers called on Jokowi to invoke the strongest possible moratorium on burning on all peatlands and to revoke peatland licenses of concessions held by companies that have not yet been used or planted.[84] The fires were raging out of control in late October 2015 while Jokowi was on an official visit to the United States. In a meeting with President Barack Obama during which the haze was discussed, Jokowi declared, "We have peat fires, and the effort to extinguish it is quite challenging."[85] In recognition of the severity of the situation, Jokowi cut his trip to the United States short to return to Indonesia, and on return visited the affected areas. Indonesia had already deployed 10,000 troops to combat the fires and he ordered additional measures.

In an attempt to exert additional pressure on Indonesia, ASEAN environment ministers met in late October 2015, expressing their concern over the fires in Sumatra and Kalimantan. The ministers reviewed regional cooperation measures taken under the ASEAN Agreement on Transboundary Haze Pollution, which was signed by all member states in 2003, except Indonesia. Indonesia finally ratified it in 2014 under President Yudhoyono. The agreement calls on countries to take actions to prevent, monitor, and mitigate the haze by controlling the source of fires, exchanging information and technology, and cooperating with one another to manage outbreaks. Neighboring states had long questioned whether Indonesia's ascension to the agreement would lead to change on the ground. Signatory countries agree to joint monitoring systems for locating fire hotspots, but governments (particularly Indonesia's) are reluctant to share map data, and there are no provisions in the treaty for enforcement or penalties when states fail to take preventive action on forest fires.[86]

At their October 2015 meeting, the environment ministers expressed concern and sympathy for those affected by the haze, agreed to share information and cooperate to fight fires, revised the Joint Emergency Response Agreement to Operationalize the Alert Levels and Trigger Points and Actions, and noted that most member states had contributed to the ASEAN Transboundary Haze Pollution Control Fund, which stood at $50 million.[87] These remarks are consistent with earlier regional discussions and efforts to address environmental and climate-related issues. On the surface, there is a fair amount of consensus within ASEAN about the need for collective action to address the fires.

Backing up cooperative words with constructive collective action, however, is a different story. In 2012, Singapore proposed and attempted to develop an ASEAN Haze Monitoring System to monitor hotspots and identify errant companies engaging in irresponsible practices to clear land. To date, however, none of its neighbors have provided maps to Singapore, illustrating how national and regional interests conflict, and make the implementation of the ASEAN Agreement on Transboundary Haze Pollution difficult.

Jokowi Responds to the Haze with Strong Regulations

At the Paris climate change meeting in December 2015, the haze was a key topic of discussion and Jokowi promised new measures to combat it.[88] In 2016, the Jokowi administration took a number of actions to address deforestation, including many steps to close loopholes in earlier regulations. In April, Jokowi announced a plan to ban new palm oil and mining permits. He framed the licensing freeze as a continuation of the 2011 moratorium that banned new permits to develop primary forests and peatlands. In a statement designed to head off protests from oil palm plantation owners, Jokowi stated that "current plantations are enough, as long as the seeds are proper, it is possible to increase productive by two times."[89]

In May 2016, the Environment and Forestry Ministry, which must approve the release of land, moved to implement the ban. In the first round of a four-phase review, the ministry rejected all outstanding requests to plant oil palms, which one estimate claims saved 851,000 hectares or 3,300 square miles from conversion. In the second round, the ministry canceled a number of in-principle approvals it had already issued to prospective oil palm developers, taking another 600,000 hectares off the table. After a company obtains an in-principle approval, the ministry must sign off on the delineation of the land in question before it issues a forest release permit to the company. The forest release permits were the focus of the third phase of the review, and in July 2016 Minister Nurbaya announced the cancellation of all forest release permits for the 2015–2016 period. She claimed that the initial findings of the third round provided "a strong legal basis for doing this."[90] In the fourth round, the ministry will review all outstanding forest release permits.

The Inpress will also close loopholes that weakened earlier efforts to avoid deforestation. Critically, palm oil and forestry resource companies will be forbidden from starting any activities on their idle peatland concessions. This significantly strengthens the original moratorium, which banned permits only on land that had not been converted into concessions. Now, land on concessions that have been converted to plantation use, but not developed, will be banned. This is a significant shift. Combined with the revocation of the in-principle permits like those awarded in 2010 and 2011 in the run-up to the moratorium, it is a significant attempt to redress some of the weaknesses of earlier regulations.[91]

In another move to strengthen oversight and control of peatlands, the government created a Peatland Restoration Agency (BRG) in February 2016, which is charged with monitoring and restoring peatlands. According to the BRG, the Inpres will significantly increase the amount of peatland under the moratorium. The BRG estimates that of Indonesia's 20.6 million hectares of peatland, 8.4 million are currently covered by the moratorium and the new prohibition will add another 4.4 million, making over half of the country's peatland exempt from development.[92] Tracking peatland use requires cooperation from other agencies. The critical role peat plays in Indonesia's carbon profile and the haze meant that the creation of the BRG was welcomed by the international community. Norway offered $25 million of support to the BRG.

Beyond monitoring, a key goal of the BRG is to restore peatlands. Peatlands are naturally swampy and burn only when they are drained of water for development. Jokowi has set a target of 2 million hectares of peatlands for restoration by 2019. According to the head of BRG, Naizr Foead, a former environmental activist, restoration work requires the "urgent and constructive cooperation with business."[93] Plantation firms whose land falls into the 2.2 million hectares of peatland that the BRG has identified as requiring restoration will be required to undertake the restoration on their own concessions with their own resources. In this way, not only does the government obtain assistance in reducing the potential for forest fires, but it also raises the cost to the plantation owner. Before, plantation owners could earn profits while others paid the economic, health, and environmental costs of their activities. Now that the private sector must bear these costs, they may bring their activities into line with the social good, ultimately reducing GHGs.

Holding Private Actors Accountable

Indonesia has also been using the courts to sue concession holders and hold them accountable for their role in the haze. The Environment and

Forestry Ministry's director general for law enforcement, Rasio Ridho Sani, stated in March 2016 that the government had taken ten companies to court and would bring suit against nine more because they had caused the fires by failing to carry out their responsibility as stipulated in their permits.[94] In addition, he stated that twenty-three other companies had been penalized for the fires, including sixteen that had their licenses suspended and three that have been permanently banned from operating in Indonesia. According to Rasio, the ministry has been suing plantation firms for causing fires since 2013, but proving culpability was difficult unless the perpetrators who started the fires were caught red-handed.[95] Indeed, the lack of concrete evidence was cited by the Palembang district court as the reason it ruled against the government in its 7.8 trillion rupiah ($570 million) suit against pulpwood plantation company Bumi Mekar Hijau for clearing land in 2014 by using the outlawed slash-and-burn technique.[96] The ministry appealed the decision, but the difficulty of proving culpability—something the One Map should strengthen—gives Indonesia's often corrupt legal system reasonable doubt to act in the interests of powerful private actors rather than the public good.

Singapore's Unilateral Legal Efforts to Combat the Haze

In an effort to overcome the lack of regional cooperation and enforcement mechanisms unilaterally, Singapore's parliament passed the Transboundary Haze Pollution Act (THPA) in 2014 that enables its regulators and courts to sue foreign individuals or companies that cause severe air pollution in Singapore. This law is an attempt by Singapore to hold those who create the haze accountable for the enormous social costs their activities impose on others. The law is also a unilateral attempt by Singapore to deter the fires after multilateral efforts have failed. If the companies and their directors had to pay the costs, then they might factor those into their decisionmaking up front. In the language of public goods theory, it is an attempt to align the public and private costs.

There are technical and political complexities involved in using this tool. Technically, holding private companies and their directors responsible for pollution requires, first that the precise location of the fire can be found and, second, that the holder of the concession can be identified. Both are problematic. Pinpointing the precise location of peat fires that burn underground can be extremely problematic, particularly during the haze itself. Furthermore, Indonesia's overlapping concession maps often make it difficult to determine the company that has the right to exploit specific forest areas and, as discussed earlier, this is a key obstacle that has hindered Indonesia's own law enforcement efforts.

The corporate and NGO members of the Roundtable on Sustainable Palm Oil, the world's largest association for the ethical production of palm oil, agreed to publish its growers' maps in the name of transparency and good governance. However, they have not been published at the time of this writing because growers in Malaysia and Indonesia claim that their governments will not permit them to release the maps. Indonesia's land minister, Ferry Mursyidan Baldan, issued a letter in May 2016 stating that companies were free to release their maps since doing so violated no law.[97] But illustrating the lack of coordination between Indonesian ministries that complicates environmental governance, companies pointed to a letter signed by Gamal Nasir, the Agriculture Ministry's plantation director, which stated that releasing concession maps was illegal. Until that letter was revoked and clarity over government regulation was established, companies argued that they could not release their maps for fear of prosecution.

Beyond the technical issues, however, are the political obstacles to Singapore's efforts to use its new law. In 2016, Singapore issued notice to six Indonesian-based companies requiring them to explain how they planned to prevent a recurrence of the devastating 2015 fires. Only two companies responded.[98] Singapore's National Environment Agency (NEA) then obtained a court warrant for an executive from one of the four companies who failed to respond. According to the NEA, if the executive enters Singapore, he can be detained for the purpose of investigation.

Indonesia's strong attachment to the principle of sovereignty means that it has always opposed the extraterritoriality nature of the THPA. The Foreign Ministry "strongly protested" Singapore's threat to detain the executive.[99] Indonesian environment and forestry minister Siti Nurbaya Bakar claimed that the multilateral ASEAN Agreement on Transboundary Haze Pollution was the proper instrument for regional cooperation, and that Singapore's use of the THPA failed to demonstrate "mutual respect" to Indonesia, a sentiment echoed by Vice President Jusuf Kalla.[100] Singapore officials have repeatedly stated that their efforts are not directed at Indonesia specifically and that its laws do not violate Indonesian sovereignty. Instead, the drivers of the haze were commercial entities whose blatant disregard for the social consequences of their actions should not go unpunished.[101] Referring to the THPA, a spokesperson for Singapore's Ministry of Environment and Water Resources (MEWR) stated, "Indonesia should welcome this tool as a new and additional tool to curb irresponsible activities that have impacted the health, social, and economic well-being of Indonesians and people of the region."[102] Rather than view the THPA as a tool for bilateral collaborative efforts, however, Indonesian officials perceived it an affront to Indonesian sovereignty and a source of conflict. In a tit-for-tat move, Minister Nurbaya announced that she had launched a review of all bilateral haze agreements with Singapore and threatened to cancel bilateral agreements between the two countries. Moreover, she also instructed local governments to refrain from cooperation with Singapore for "the time being."[103] Minister Nurbaya's decision is reminiscent of previous occasions when the haze triggered verbal tit-for-tat comments by Indonesian leaders. In 2015 Vice President Kalla responded to criticism of the haze by asking Indonesia's neighbors to be grateful for the clean air they had for eleven months of the year: "For eleven months they had nice air from Indonesia and they never thanked us."[104]

Conclusion

Climate change is already negatively affecting Southeast Asia, and future problems relating to food and water security could be acute. Deforestation is something that countries could take concrete action to curtail and which would have immediate positive effects locally for communities that receive financial incentives to conserve forests instead of cutting them down and regionally as people would be spared from the haze. Indonesia's experiences with deforestation efforts through REDD+ demonstrate the difficulty in putting cooperative rational solutions into action in countries with weak technical capacity, weak governance, and powerful private actors pushing back against public policy that would negatively impact their private interests. The multitude of actors whose cooperation is required to mitigate GHGs illustrates the challenges of improving cooperation and compliance when governance is extremely fragmented.

Indonesia has taken significant steps to mitigate climate change since 2010 to set the country on a course that will shift its current high-carbon development model into a low-carbon sustainable one. Technically, Indonesia's progress toward its One Map and the development of other tools to monitor its forests constitute a large and significant step. Organizationally, Indonesia has rationalized its forestry governance and created new institutions like the BRG to monitor peatlands. It has extended the moratorium, broadened its scope, and closed loopholes that enabled businesses and corrupt public officials to pursue their private interests at the expense of the broader public one. Legally, Indonesia has begun to hold those who cause the haze accountable. All of these are moves in the right direction.

Nevertheless, much more needs to be done and significant obstacles still exist. Moratoriums are not an end in itself, but an attempt to halt the current pace of deforestation to provide an opportunity to plan for better forest governance through the institutionalization of better policy processes, better data collection, and perhaps new regulations. As the debate and weak implementation over the moratorium illustrate, such steps are highly contested, difficult to enforce, and resistant to easy solutions. Effective cooperation on climate change problems would require better vertical government coordination between national-level agencies in Jakarta and provincial and local levels of government. It would require political leaders to stand up to powerful business interests who have close ties to political officials at all levels of government and to be able to confront and deter corruption. The response of the Jokowi administration to the 2015 haze demonstrates an intent to do so, but these steps will need to be sustained.

At the regional level, ASEAN has provided a forum for discussion and for the codification of an agreement specifically targeting transboundary haze, including creating a fund to help countries pay for monitoring and combating potential hotspots and fires. However, such agreements are going to be successful only if member states want them to be. Indonesia believes that its size and history give it the authority to serve as political leader in ASEAN but, on the specific issue of haze, leadership requires standing up to powerful domestic actors. Effective regional cooperation also requires recognition that actions in one country negatively impact another and necessitate working with regional partners to address them. Instead, the haze has triggered nationalist reactions from key Indonesian leaders and conflict.

The 2015 haze may well turn out to be a critical juncture in Indonesian efforts to mitigate climate change. To date, however, issues with technical monitoring and verification mechanisms, coordination problems between different government agencies and levels of government, and political contestation over policy between those who want to reduce Indonesia's carbon footprint and those who benefit from the current system have all combined to frustrate Indonesian efforts to curb deforestation.

Notes

1. US Environmental Protection Agency, "Climate Change Basics," Washington, DC, information snapshot as of January 19, 2017, https://19january2017snapshot.epa.gov/climatechange/climate-change-basic-information_.html.

2. Bernadinus Steni, "Climate Change and ASEAN," *Jakarta Post,* May 10, 2011.

3. United Nations, "The Paris Agreement: Status of Ratification," United Nations Framework Convention on Climate Change, November 4, 2016, http://unfccc.int/paris_agreement/items/9485.php.

4. Sureeratna Lakanavichian, "Trends in Forest Ownership, Forest Resource Tenure and Institutional Arrangements: Are They Contributing to Better Forestry Management and Poverty Reduction? A Case Study from Thailand," FAO, February 2006, www.fao.org/forestry/10809-09f8870885bd8d85106e0a87cd906b784.pdf; Zelda Soriano, "Climate Change Demands that ASEAN's Future Is Cooperative," *The Nation,* May 1, 2013.

5. Daniel R. Richards and Daniel A. Friess, "Rates and Drivers of Mangrove Deforestation in Southeast Asia, 2000–2012," *Proceedings of the National Academy of Science of the United States of America* 113, no. 2 (2016): 344–349; John Vidal, "Greater Mekong Countries 'Lost One-third of Forest Cover in 40 Years,'" *The Guardian,* May 2, 2013.

6. World Bank, "Indonesia and Climate Change, Executive Summary," Jakarta, May 2007, 3.

7. US Department of State, "Indonesia: Background Note," January 17, 2017, www.state.gov/r/pa/ei/bgn/2748.htm.

8. Frank Jotzo, "Indonesia's Role in International Climate Change Policy," *East Asia Forum*, October 11, 2011.

9. World Bank, "Indonesia and Climate Change: Current Status and Policies, Executive Summary," 2007, 5.

10. "The Key Players in Climate Change," *New York Times,* August 11, 2016.

11. Ibid.

12. "Letter of Intent Between the Government of the Kingdom of Norway and the Government of the Republic of Indonesia on Cooperation on Reducing Greenhouse Gas Emissions from Deforestation and Forest Degradation," Royal Norwegian Embassy in Jakarta, May 26, 2010, www.norway.or.id/?querystring=LOI+Redd (emphasis added).

13. T. Sima Gunawan, "Questions Loom over 1-Billion Tree Planting Program," *Jakarta Post,* June 5, 2010.

14. "SBY Vows to Protect Indonesia's Rainforests," *Jakarta Globe,* September 27, 2011.

15. "Indonesia Sets Aside 45% of Forest-Rich Kalimantan to Be the World's Lungs," *Forests News Blog,* Center for International Forestry Research (CIFOR), January 25, 2012, http://blog.cifor.org/7144/indonesia-sets-aside-45-of-forest-rich-kalimantan-to-be-worlds-lungs/.

16. Peter Riggs, "Indonesia's Colossal, Vital and Unusual Role in the Global Climate Change Debate," paper presented at the USINDO Open Forum, Washington, DC, May 4, 2010.

17. Ibid.

18. Center for International Forestry Research (CIFOR), *Simply REDD: CIFOR's Guide to Forests, Climate Change, and REDD,* 2009, 5, https://www.cifor.org/library/2812/simply-redd-cifors-guide-to-forests-climate-change-and-redd.

19. Ibid.

20. "SBY Vows to Protect Indonesia's Rainforests."

21. Graham Lloyd, "Loggers Turned Conservationists," *The Australian,* October 1, 2011.

22. Sunagh Creah, "Corruption Could Jeopardize Indonesia's Climate Change Deals," *Jakarta Globe,* September 17, 2010.

23. *Simply REDD.*

24. Many observers claim that the Norway deal has effectively transformed Susilo Bambang Yudhoyono's 26 percent pledge into a 41 percent commitment.

25. Faisal Maliki Baskoro, "Nestle Shows Its Sweet Side to Greenpeace, Sinar Mas in Indonesian Palm Oil Fight," *Jakarta Globe,* May 17, 2010.

26. Lloyd, "Loggers Turned Conservationists."
27. Ibid.
28. Rhett Butler, "Indonesia Grants Exemptions from Logging Moratorium for 2.6m ha of Forest," *Mongobay,* December 21, 2011.
29. "Letter of Intent Between the Government of the Kingdom of Norway and the Government of the Republic of Indonesia."
30. Ibid., 2.
31. Ibid., 3.
32. Michele Kovacevic, "Norwegian Minister for Environment Praises Indonesia's Fight Against Climate Change," *Forests News Blog,* CIFOR, September 27, 2011, http://blog.cifor.org/4260/norweigian-minister-for-environment-praises-indonesias-fight-against-climate-change/.
33. Frances Seymour, Nancy Birdsell, and William Savedoff, "The Indonesia-Norway REDD+ Agreement: A Glass Half Full," Policy Paper No. 56 (Washington, DC: Center for Global Development, February 2015).
34. "Definitions: Indicative Definitions Taken from the Report of the Ad Hoc Technical Expert Group on Forest Biological Diversity," United Nations Convention on Biological Diversity, 1992, www.cbd.int/forest/definitions.shtml.
35. "Preventing the Risk of Corruption in REDD+ in Indonesia," Working Paper No. 80 (Bogor, Indonesia: CIFOR, 2011).
36. Ibid., 4.
37. "Indonesia's Forest Moratorium: A Stepping Stone to Better Forest Governance?" Working Paper No. 76 (Bogor, Indonesia: CIFOR, 2011), 5.
38. Kara Moses, "Lack of Forest Definition 'Major Obstacle' in Fight to Protect Rainforests," *The Ecologist,* October 20, 2010.
39. Ibid.
40. David Fogarty, "How Indonesia Crippled Its Own Climate Scheme," Reuters, August 16, 2011.
41. Ibid.
42. Ibid.
43. "Preventing the Risk of Corruption in REDD+ in Indonesia," 14.
44. "SBY Vows to Protect Indonesia's Rainforests."
45. Aditya Suharmoko, "SBY Vows Direct Oversight of Norway-Financed Projects," *Jakarta Post,* May 30, 2010.
46. Philip Jacobson, "Nine Months In, How Has Jokowi Fared on the Environment?" Mongabay, July 23, 2015, https://news.mongabay.com.
47. Ibid.
48. Olivia Rondonuwu, "Indonesia Divided over Forest Moratorium," Reuters, January 5, 2011; "Indonesia Palm Oil Industry Market Drivers, Challenges and Competitive Landscape Analysis," MarketersMedia press release, January 20, 2016.
49. Agus Purnomo, "Environment Caught in the Extremes," *Jakarta Globe,* April 19, 2011.
50. Member of the REDD+ Task Force, interviewed by Ann Marie Murphy, Jakarta, January 8, 2012.
51. Jacobson, "Nine Months In, How Has Jokowi Fared on the Environment?"
52. "Interview with Teguh Surya, WAHLI: We Are Against REDD, We Are Against Carbon Trading," REDD-Monitor, March 9, 2012, http://www.redd-monitor.org/2012/03/09/interview-with-teguh-surya-walhi-we-are-against-redd-we-are-against-carbon-trading/.
53. Yuni Saloh, "Moratorium Isn't the End of Business," *Jakarta Globe,* July 14, 2011.

54. Zubiada Nazeer, "Indonesia Loses Forests as Politicians Bicker," *The Straits Times,* April 19, 2011.

55. Mustaqim Adamrah, "RI Risks Business, Forests with US$1b Forest Moratorium Delay," *Jakarta Post,* January 6, 2011.

56. Rondonuwu, "Indonesia Divided over Forest Moratorium."

57. Ibid.

58. Augus Purnomo, President Yudhoyono's former chief climate change adviser and a member of Indonesia's National Council on Climate Change, interviewed by Ann Marie Murphy, Jakarta, January 20, 2012.

59. Adianto P. Simora, "Govt Torn Between Courting Businesses, Protecting Forests," *Jakarta Post,* May 24, 2011.

60. "Indonesia's Forest Moratorium: A Stepping Stone to Better Forest Governance?" 2.

61. Ibid., 6.

62. Ibid., 2.

63. Ibid., 3.

64. Ibid., 6.

65. Ibid., 7.

66. Simamora, "Govt Torn Between Courting Business, Protecting Forests."

67. Adianto P. Simora, "Govt Bows Down to Planters, Forest Decree Means Little," *Jakarta Post,* May 21, 2011.

68. "Indonesia's Forest Moratorium," 7.

69. Simora, "Govt Bows Down to Planters, Forest Decree Means Little."

70. Government official committed to REDD who previously worked for an environmental NGO, interviewed by Ann Marie Murphy, Jakarta, January 20, 2011.

71. "Forest Moratorium to Cut Indonesian Crude Palm Oil Production," Dow Jones, June 21, 2011.

72. Angela Dewan, "Indonesia Delays Forest Ban," *Asia Sentinel,* April 13, 2011.

73. "Preventing the Risk of Corruption in REDD+ in Indonesia," 19.

74. Simora, "Govt Bows Down to Planters, Forest Decree Means Little."

75. Rondonuwu, "Indonesia Divided over Forest Moratorium."

76. Angela Dewan, "New Tech, Better Map on Tap to Protect Indonesian Forests," *Forests News Blogs,* CIFOR, May 19, 2014, http://blog.cifor.org/22534/new-tech-better-map-on-tap-to-protect-indonesian-forests?fnl=en.

77. Nanang Indra Kurniawan, "Getting the One Map Policy Right," Indonesia at Melbourne, May 11, 2016, http://indonesiaatmelbourne.unimelb.edu.au/getting-one-map-policy-right/.

78. Ibid.

79. Euan McKirdy, "Southeast Asia's Haze Crisis: A 'Crime Against Humanity,'" CNN, October 29, 2015.

80. Ibid.

81. Kate Lamb, "Indonesia's Fires Labelled a 'Crime Against Humanity' as 500,000 Suffer," *The Guardian*, October 26, 2015.

82. World Bank, "Indonesia's Fire and Haze Crisis," August 21, 2016, www.worldbank.org/en/news/feature/2015/12/01/indonesias-fire-and-haze-crisis.

83. "Singapore's GDP Will Take a Hit from the Haze as Countries Issue Travel Warnings," *The Straits Times,* October 7, 2015.

84. "Under-Fire Jokowi Prepares Biggest Shift Yet in Indonesian Haze," *Mongabay Haze Beat,* October 26, 2015, http://news.mongabay.com/2015.

85. Juliet Eilperin and Chris Mooney, "Why the 'Biggest Climate Story on the Planet' Is Happening in Indonesia Right Now," *Washington Post,* October 26, 2015.

86. Kanis Dursin, "ASEAN Agreement on Haze? As Clear as Smoke," Inter Press Service, October 10, 2015.
87. "Land-Use Fires and Smoke Pollution in Indonesia," Global Fire Monitoring Center, November 18, 2015, www.uni-freiburg.de/GFMCnew.
88. "Under-Fire Jokowi Prepares Biggest Shift Yet in Indonesian Haze."
89. Fidelis E. Satriastanti, "Jokowi Bans New Oil Palm and Mining Concessions," Eco-Business, April 16, 2016, www.eco-business.com/news/jokowi-bans-new-oil-palm-and-mining-concessions/.
90. Philip Jacobson, "Indonesia's Palm Oil Permit Moratorium to Last Five Years," Mongabay, August 15, 2016, https://news.mongabay.com/2016/07/indonesias-palm-oil-permit-moratorium-to-last-five-years/.
91. Hans Nicholas Jong, "Govt to Put Moratorium on Firms with Idle Concessions," *Jakarta Post,* June 18, 2016.
92. Ibid.
93. Ibid.
94. Arlina Arshad, "Jakarta to Sue 9 More Firms over Forest Fires Linked to Haze," *The Straits Times,* March 1, 2016.
95. Ibid.
96. Jong, "Govt to Put Moratorium on Firms with Idle Concessions."
97. Philip Jacobson, "Malaysian Palm Oil Companies Say Their Concession Maps Are State Secrets," Mongabay, May 6, 2016, https://news.mongabay.com/2016/05/malaysian-palm-oil-companies-say-concession-maps-state-secrets/.
98. Philip Jacobson, "Indonesia Jostle over Anti-Haze Measure," Mongabay, August 9, 2016, https://news.mongabay.com/2016/05/singapore-indonesia-jostle-anti-haze-measures/.
99. Ibid.
100. Francis Chan, "Singapore Targeting Private Firms Linked to Indonesian Fires," *Jakarta Post,* June 16, 2016.
101. Ibid.
102. Ibid.
103. Jacobson, "Indonesia Jostle over Anti-Haze Measure."
104. "Indonesia's Vice-President Jusuf Kalla Criticizes Neighbours for Grumbling About Haze," *The Straits Times,* March 5, 2015.

3

Food Security: Beggar Thy Neighbor?

As Chapter 1 lays out, enormous gains in life expectancy and our increasing ability to grow more crops under diverse climate conditions may run up against even tougher pressures in the future. The World Health Organization defines "food security" as existing "when all people at all times have access to sufficient, safe, nutritious food to maintain a healthy and active life."[1] Food security is built on three pillars: *food availability,* consistently available quantities of food; *food access,* having sufficient resources to obtain appropriate food for a nutritious diet; affordability; and *food use,* appropriate use based on knowledge of basic nutrition and care, adequate water, and sanitation.[2] In 2017, we are looking ahead to a future of possible widespread food insecurity, especially in the developing world.

Southeast Asia in particular is facing escalating food challenges. Countries in the region continue to enjoy sustained economic growth despite the Chinese, European, and US slowdown. And countries continue to reduce overall levels of poverty. Yet over 15 percent of Asia's undernourished population lives in Southeast Asia. The region has abundant natural resources and is rich agriculturally, but environmental stresses and changing demographics, lifestyles, and eating habits threaten many key ecosystems and thus pose a threat to future food production. Historically, food shortages have been a result of poor social, economic, and political policies, often combined with environmental problems such as drought, floods, and blight that can exacerbate unequal allocation of resources. Food shortages are thus a product of failures of availability. While most people do not see governments as being responsible for directly supplying food to citizens, governments do create the conditions necessary for markets to function properly. Governments impact food availability and prices by enacting policies such as subsidies to producers and to poor consumers, taxing some goods and not others, or buying up quantities of a crop to provide stability in a market. Today, the most probable cause of food insecurity for millions of people around the world will be changing demographics coupled with the effects of climate change, and the inability of many governments to allocate resources appropriately to feed the largest number of people. Southeast Asia is faced with all of these challenges. Some important actions are currently being taken to address issues of food security, but adoption and implementation has been slow and uneven. In this book, we chose to look at the rice trade specifically as an example of food insecurity. Our decision to use the rice market in Southeast Asia as a lens for understanding access to food stems from the fact that rice is still an important staple food crop for communities in the region and, because some countries are exporters and some are importers of rice, it demonstrates the need for states to work together to ensure adequate food supplies for the greatest number of people. There is tremendous regional diversity in Southeast Asia; it is home to some of the world's largest rice exporters, but also home to some of the largest importers and largest consumer populations in terms of rice consumption per capita. In this chapter, we look at the creation of institutions in Southeast Asia to promote information sharing and policy coordination on rice cultivation and trade. We also examine some individual nations' policies and their choices and challenges.

There have been some real and promising efforts to cooperate regionally on food security; however, these measures are far from being firmly instituted and operational. Food security is a critical issue for human security; citizens need access to functioning markets to acquire the food necessary to sustain life, but food security is also related to regime security. While there is little likelihood that one country is going to invade another country over food prices and trade conditions, governments are highly aware of food prices and supplies because of the possible link between food insecurity and political instability. Governments want to maintain power, and they fear the kinds of mass demonstrations and instability that can be triggered by food crises. Thus, policies surrounding access to food are an example of comprehensive security. Governments in Southeast Asia have overseen significant improvements in tackling food insecurity, yet pockets of malnourishment exist and millions of residents are susceptible to periods of food insecurity. Many governments in Southeast Asia are actively involved in agricultural production and distribution. Indonesia's government agency that oversees food distribution and price control is Bulog. Bulog does not give food directly to citizens, but it acts as a middleman between producers and consumers to ensure ample food supplies and pricing. Other countries in Southeast Asia have similar agencies. Thus, governments have a significant impact on food availability and pricing.

Access to food is a basic human need and we take for granted that states will help create conditions whereby food is able to be grown, produced, brought to market, imported, or otherwise made available to citizens for consumption. There are times when we are starkly reminded how difficult this can become. In the wake of a natural disaster a state's ability to ensure access to food becomes critical but, even under normal circumstances, state capacity to ensure food security is a critical public good. For states that rely on importing food commodities, trading conditions for crops become an essential security element. Thus, exporters and importers and governments need to work across borders to secure food access to the greatest number of people. However, such cooperation may go against entrenched and powerful interests within countries. The problem of food security is clearly transboundary in nature because what impacts farmers in one country (e.g., the low crop yields in 2016 in Vietnam because of drought) will have spillover effects elsewhere. Poor Filipinos, for whom rice is a key component of their diet, will find it harder to afford and buy rice as supplies shrink and prices increase. Often, the interests of importers and exports are at odds; yet if all citizens are to have access to food, there needs to be coordination between importing and exporting countries. In this chapter, we look at efforts to address these fragmented and cross-cutting interests in the Southeast Asian rice market. We look at national responses to the rice "crisis" in 2007, at the role of state bureaucracies, and at interest groups in key rice-producing countries as well as at regional and international institutions to understand cooperative or uncooperative behavior. In the wake of this crisis, there have been efforts to improve cooperative efforts to prevent future panics and price hikes for this crucial crop.

In response to food challenges of the past, countries in Southeast Asia have implemented several measures to address the risks of food insecurity:

> These include export restrictions, price controls, price subsidies, and import facilitation. Such approaches are understandable and at times seemingly the only option available to governments. However, there are both costs and benefits from these types of state intervention in food markets, as these strategies potentially involve competing objectives, that of protecting consumers, against that of assisting agricultural producers to benefit from rising prices.[3]

These measures (undertaken by a number of countries in Southeast Asia at one point in time or another) constitute beggar-thy-neighbor policies where countries enact measures to protect their own citizens. The larger effect of these policies can sometimes be a worsening of conditions (in this case, food security conditions) across borders.

While these actions are at times necessary and are understandable, they again illustrate the fundamental problem of international relations. What one country does (internally) then impacts other countries. When a food exporter decides to reduce or stop exporting that product, it can have dire consequences elsewhere for those who rely on importing that food commodity. Thus, states need to cooperate with each other, but often find it difficult to do so. Cooperation is hindered by imperfect information, a lack of regular mechanisms for designing agreements, and variation in interests being favored from states. This illustrates fragmented governance and the problems of complexity in governance architecture.

Food Insecurity

Many factors and trends are emerging that threaten Asian and global food security. While strong economic growth has bumped a number of states toward middle-income status, a large number of the world's undernourished live in Asia (India accounts for 43 percent of the undernourished, and China for 24 percent). The region's population is projected to increase from 3.6 billion to 4.5 billion people. Most of that growth will occur in urban areas, as the urban population is set to surpass the rural population as early as 2028.[4] At the same time, there has been a global decline in investment in agriculture—from a 20 percent share of ODA for agriculture in 1979 to 5 percent in 2007—and it continues to drop.[5] There is also a paradox: economic growth and prosperity are unquestioningly a good thing; however, as societies become wealthier, they also eat differently, consuming more animal products. Thus, there is increased pressure on agriculture to produce feed for livestock instead of food for human consumption. This can result in increased prices for grains and staple commodities, further hurting poorer citizens and areas.

Another complicating factor making food security difficult to tackle is that there is tremendous variation across borders in Southeast Asia as well as within countries. Not surprisingly, poorer countries in the region (e.g., Laos, Cambodia, Timor Leste, and Myanmar) have the highest levels of food insecurity; this is even the case with Cambodia, and to a lesser extent Myanmar, as rice exporters. However, within some of the better-off countries in Southeast Asia, there are significant numbers of food insecure people. Pockets of food insecurity are found in specific provinces or states, and so it is important to map in-country food insecurity as well as national and regional trends. Particularly vulnerable are poor landless families or communities (often this means the urban poor), female-headed families, and rural landless households.[6] Also vulnerable are many ethnic minorities living throughout Southeast Asia. Swidden agriculture, sometimes labeled "slash-and-burn" practices, has come under scrutiny and criticism for the environmental damage that it causes. Since the mid–twentieth century, swiddeners' everyday lives have been radically transformed. Once, communities as diverse as the Dayaks in Indonesia, Hmong in Cambodia and Laos, and the Orang Asli in Malaysia, all practiced swidden. Forestland would be burned and debris largely removed. Ash would be left in place and the land would be planted with crops. Once harvested, the land would then be allowed to remain fallow to regenerate or replenish the soil. Forestland would grow back over several years and then the cycle would begin again. Such practices have been criticized and often outlawed because of the release of greenhouse gases into the atmosphere from burning.

However, what has taken the place of such traditional practices is far worse. States in the region have come to try to manage forest and agricultural lands, regulating what land can be used for what purposes. This has often had the effect of forcing resettlement of these communities. But sometimes the opposite has happened: transmigration programs, like in Indonesia, have resulted in some ethnic groups becoming overwhelmed by resettled populations in their midst, thus transforming the indigenous community into a minority in their own homeland. Nations have increasingly asserted control over swidden lands and the people who farmed them, often completely or partially dispossessing those who depended on the practice.[7] While swidden is viewed as environmentally harmful, it is nowhere near as destructive as the industrial logging, plantation expansion for the palm oil industry, and the massive deforestation that has taken place in its stead. Some of the ethnic communities marginalized by changes in rural land use are among the poorest and most food insecure in Southeast Asia.

Climate Change and Its Impact on Food Security

Climate change is projected to have significant effects on Southeast Asia's agriculture. Climate change affects agricultural and food production in complex ways. It affects food production directly through changes in ecological conditions and indirectly by affecting growth and distribution of incomes, living patterns, and demand for agricultural output. Changes in temperature and precipitation associated with climate change and continued emissions of greenhouse gases will bring changes in land suitability and crop yields. Global and regional weather conditions are also expected to become more variable, with increases in the frequency and severity of extreme events such as cyclones, floods, hailstorms, and droughts. By bringing greater fluctuations in crop yields and local food supplies and higher risks of landslides and erosion damage, they can adversely affect the stability of food supplies and thus food security. The 2015–2016 severe El Niño season produced unusually significant drought conditions, which, as discussed in Chapter 4 on water security, adversely impacted crop yields in Vietnam, Cambodia, Laos, and Thailand. Since Vietnam, Thailand, and Cambodia are rice exporters, this has had an adverse effect on food availability in countries like the Philippines and Malaysia.

Another source of food insecurity comes from biofuel production. Many questions have been raised over the ethics and long-term feasibility of the conversion of agricultural production for food (for human consumption and for feeding livestock) to the use of agriculture to produce fuel. While high energy prices and energy concerns of the developed world led to an increase in the use of biofuels, balancing food and energy needs is a major issue for both developed and developing nations. Increases in biofuel production are causally related to price surges in agricultural commodities. This means that countries that are already prone to food shortages are more exposed to food insecurity and related problems.

Biofuels

Production of biofuels affects all three pillars of food security: availability, affordability, and accessibility. As more land is converted to use for producing biofuels, the availability of food crops is reduced. Price surges in commodities stemming from the subsequent high demand are likely to affect the poorest populations most severely. And this land conversion can also result in a concentration of ownership that could drive the world's poorest farmers off their land and into deeper poverty.[8] This situation pits developed countries whose interests tend to favor policies that will stabilize energy supplies and prices against developing countries where food insecurity is a more pressing problem. Countries in Southeast Asia are impacted differently by the use and demand for biofuels. Indonesia and Malaysia have a market advantage on this issue, based on the high quality of fuel that can be produced from palm oil. Palm oil is one of the most efficient biofuels and it impinges least on food security. However, intensification of palm oil production does raise serious environmental concerns, as it often results in the cutting down of rain forests to increase palm oil plantation acreage.[9]

Myanmar illustrates a critical problem posed by the increasing biofuel trend. Myanmar is pursuing policies to convert land from food production to investing heavily in jatropha cultivation for fuel production; the Philippines and India are also exploring expanding jatropha production. In Myanmar, this decision has meant land grabs and displacement for already poor and marginal populations.[10] Global trends in worldwide biofuel production raise questions of sustainability and environmental impact. Continued use of biofuels threatens to decrease food security in producing countries, especially poorer ones where food shortages occur frequently, and such problems will only be exacerbated by climate changes. If developing countries could use the export of biofuels to gain market advantages and significantly increase economic development, this ironically could pave the way for larger government outlays on a more comprehensive safety net, which would act to thwart food insecurity in large segments of their populations. However, this would take prudent and careful policymaking, and it would not happen overnight. Thus, developing countries in Southeast Asia need to be wary of jumping too fully on the biofuel bandwagon and to be mindful of not undermining goals of achieving conditions of food security.

Given the levels of urbanization, changing dietary habits, socioeconomic variation across the region, challenges of climate change, and increasing land diversion to biofuel production, Southeast Asia will need to be vigilant about food security for the foreseeable future. In 2007, we witnessed how what should have been a small shift in food-trading circumstances escalated into a regional panic. This example, detailed below, illustrates the urgent need for greater cooperation on food security throughout Asia and it points to a critical need to focus more global attention on the importance of understanding food prices and markets, not just as one economic factor but as a larger socioeconomic and political driver.

Skyrocketing Rice Prices in 2007

Although we began this chapter with the basic assumption that food insecurity is often driven by poor policies and poor growing conditions such as natural disasters or the effects of climate change, there is a recent (and critical) example of a food panic that was not triggered by either of these factors. In October 2007, the Indian government decided to buy more rice for its own food distribution programs. This was a perfectly understandable decision since the rice was intended to replace wheat, as commodity prices for wheat had escalated sharply in 2007. To make sure there was plenty of cheap rice available, the Indian government made it illegal to export most Indian rice. Global response was almost instantaneous. (See Figure 3.1.) The next day, trading prices for rice in Bangkok shot up $75 a ton, and prices continued to climb through the middle of 2008.

The jump in price triggered a worldwide panic over rising prices and possible scarcities. Individuals and governments throughout the world started stocking up on rice. During the first four months of 2008, major exporters held on to their rice crops: "Egypt, Pakistan and Vietnam all started hording their rice crop. So, there was still plenty of rice in the world, but there was less for sale."[11] The lack of availability put tremendous pressure on importing countries such as the Philippines. Rice is a large part of people's diets in the Philippines and the price spikes caused more than just anxiety—people in the Philippines were actually going hungry as rice became hard to find at any price. While the crisis might have been manifest by psychological factors, the impact was quite real. Estimates show that, before the rice price spikes in 2007, there were 542 million people in the region whose caloric intake was below the minimum dietary requirement of 1,800 calories a day. As a result of the rice

Figure 3.1 Rice Prices over Time, 2003–2014

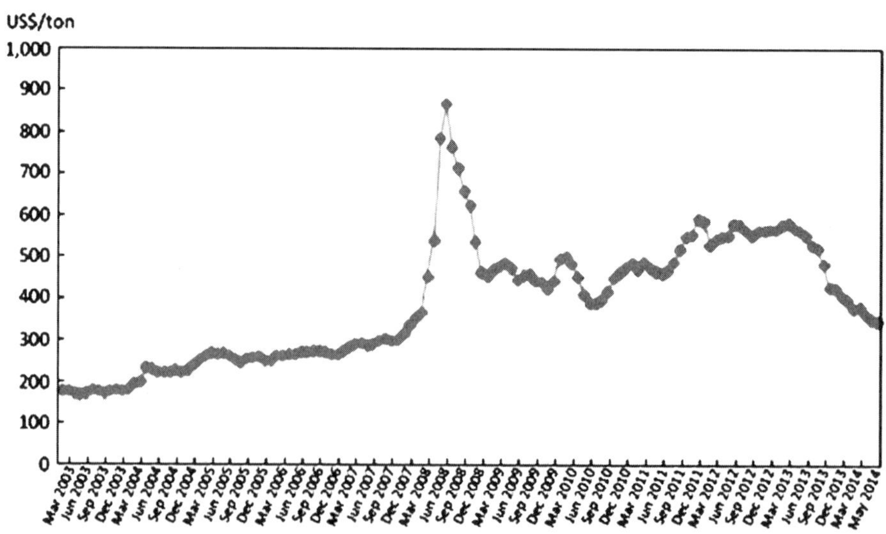

Source: International Rice Research Institute (IRRI), "IRRI—The Global Rice Market Winks at El Niño and Thai Problems," August 2014, International Rice Research Institute (IRRI), http://irri.org/rice-today/the-global-rice-market-winks-at-el-nino-and-thai-problems.

crisis, that number rose to 582 million people.[12] Another effect of the crisis was to increase the incentives for corruption. C. Peter Timmer explains that the high prices amplified the opportunities for corrupt officials to line their pockets. The Philippine government used state funds to buy rice from a state-owned company in Vietnam, Vinafood, at sky-high prices. Vinafood was able to use some of that money to buy the actual rice from Vietnamese farmers, but there still was a huge profit margin from the transaction that could then be split with officials from the Philippines government who engineered the transaction.[13]

Timmer, an economist, and Tom Slayton, a rice trader, realized that the price hikes and panic were more about psychology than actual scarcity. They knew that there were a million and a half tons of high-quality rice in Japan, WTO rice: "To settle a trade dispute with the U.S., Japan had agreed to take in a lot of American rice that it didn't really want. That rice just sits in warehouses. Japan is not allowed to export it, unless the U.S. says it can. Timmer and Slayton started a lobbying campaign. They said to the U.S., let the Japanese sell this rice just this once."[14] By mid-May 2008, the United States agreed. Just the public announcement that this rice would be coming onto world markets was enough to start bringing the price down. By the end of the year, the price of rice had been cut in half. Ultimately, none of the rice in Japan ever came on the market. The cause of the crisis was psychological and, once the fear of too little rice being available on rice markets was put to rest, prices could stabilize. The resolution of this crisis involved diverse actors: an academic, a commodities trader, US and Japanese government officials, and bureaucrats from the WTO. The hybrid nature of these actors and the ability of all parties concerned to agree on a solution is an example of how complex or hybrid systems of governance can work. But, this is one small, very specific example of a larger and more difficult puzzle over how government, nongovernment, and global organizational actors can cooperate on shared problems. The broader problem of coming up with ways to make rice markets more transparent and resilient to crises has not been solved. Diverse and fragmented interests and dynamics are harder to overcome.

What this episode clearly illustrates is that actual food supplies are just one part of the issue; availability is not just about production and amounts of a good, it is also about distribution and functioning markets. Consumers, importers, exporters (whether private companies or state agencies), and governments will act in rational ways. But this rational behavior is driven by available information (which may be imperfect) and by perception (which may not be rational). So, if a decision is made to stop exporting, because there is a belief that supplies may be low, then that decision triggers further decisions by importers, other governments, and consumers. And the results, as we saw in 2007–2008, can be quite troubling. There are solutions to this, as Timmer and others have proposed. First, countries, either individually or cooperatively, can enhance their reserves of rice and other staple crops. Second, more transparent mechanisms can be created to provide more accurate information on supply levels. The rice market in particular is not terribly transparent. If more precise information were available on supplies and prices, then decisionmaking could also be more rational and less given to panicked behavior. While these remedies may seem fairly simple, in practice they are not.

In Table 3.1, Malaysia and the Philippines are the consistently high importers of rice. Indonesia also imports large amounts of rice but given the population of Indonesia, the amounts are not as significant. Thailand and Vietnam (Table 3.2) are the largest exporters by a wide margin. It is noteworthy that levels of imports to countries did not drop precipitously in 2008 (for the year as a whole), nor did exports dip as much as we might think. There was a fall in Vietnam's exports from 5,250,000 metric tonnes to 4,735,170 metric tonnes. But, this drop is not commensurate with the panic that ensued. Thus, the problem that triggered the panic was not supply and demand per se; it was the perception of the supply.

National Concerns and Actions on Food Security

Asian countries are some of the world's largest importers and exporters of rice. On the import side, Asian countries import about 41 percent of rice traded on global markets: the Philippines imports about 3.6 percent of world rice imports, Bangladesh 3.4 percent, China 3.2 percent (but growing sharply), North Korea 2.4 percent, Japan 2.3 percent, Malaysia

Table 3.1 Import Quantities for Rice (in tonnes)

	2002	2005	2008	2011	2014
Cambodia	68,741	40,209	27,600	12,128	54,839
Indonesia	1,790,323	188,944	288,359	2,745,281	1,100,000
Malaysia	496,102	583,654	1,106,743	1,031,035	1,100,000
Philippines	1,196,159	1,821,641	2,432,001	706,471	1,850,000
Thailand	895	2,381	13,609	10,630	NA
Vietnam	40,000	340	683	2,409	2,297

Source: Data retrieved from UN Food and Agriculture Organization, "Country STAT: Official Statistics Through the CountrySTAT Food and Agriculture Data Network." Statistics compiled quarterly; figures here are through 2014. http://faostat3.fao.org/faostat-gateway/go/to/download/T/TP/E.

Note: NA = data was not available.

Table 3.2 Export Quantities for Rice (in tonnes)

	2002	2005	2008	2011	2014
Cambodia	3,846	2,318	5,536	174,045	387,061
Indonesia	3,891	42,285	1,840	803	NA
Malaysia	2,703	5,387	911	661	NA
Philippines	2	196	820	395	NA
Thailand	7,328,418	7,508,581	10,186,678	10,671,194	9,200,000
Vietnam	3,240,932	5,250,000	4,735,170	7,112,000	6,500,000

Sources: Data retrieved from UN Food and Agriculture Organization, http://faostat3.fao.org/faostat-gateway/go/to/download/T/TP/E; "CountrySTAT: Official Statistics Through the CountrySTAT Food and Agriculture Data Network." Statistics compiled quarterly; figures here are through 2014.

Note: NA = data was not available.

1.8 percent, Indonesia 1.3 percent (but variable), and Singapore 1.2 percent. On the export side, Asia farms more than 91 percent of the global rice harvest.[15] The largest rice exporters are Vietnam, Thailand, India, and Cambodia. We discuss the key countries in turn.

Malaysia

Malaysia is a rice importer. From 1996 to 2014 Malaysia on average needed to import 29 percent of its domestic consumption, a figure that has risen slightly with increased consumption over the past few years.[16] For Malaysia, rice security is a matter of national security. A July 2011 conference in Malaysia on rice intensification systems looked at how to bolster domestic rice production.[17] In reaction to global price hikes in 2007–2008, the government of Malaysia passed the Food Security Policy (FSP). Under this policy, $105 billion (506 billion ringgit) was allocated for agricultural development, particularly rice cultivation and production. In the spring of 2011, the government extended this policy. Malaysia's National Agricultural Policy (NAP) for 2011–2020 will determine whether Malaysia will be able to achieve food security through self-sufficiency, or whether Malaysia will continue to be dependent on imports. The NAP involves both input and output subsidies in agriculture, guaranteeing a minimum price for paddy output, fertilizer subsidies, and subsidies for paddy cultivation, seeds, and rice price supports. The coordinating agency for this is the Federal Agricultural Marketing Authority (FAMA), which is like Indonesia's Bulog.[18] These policies may produce short-term self-sufficiency but, if the experiences of other countries are any guide, these policies are expensive to maintain and may not be environmentally sustainable.

Indonesia

In the 1970s and 1980s, Indonesia followed policies similar to those of Malaysia and found that the cost of food self-sufficiency was extremely high. Input subsidies to help farmers, such as fuel, fertilizers, pesticides, and irrigation, were quite costly. For example, in 1986–1987, the total cost of these subsidies equaled $725 billion. The government was also heavily involved in building infrastructure for irrigation and transportation to bolster agricultural production. Although they increased food production, these programs also resulted in costly externalities. Too much fertilizer and too many pesticides were used because they were cheap to the farmers. So, not only was there a huge government outlay to increase crop yields, but there also was environmental damage from the attempt to achieve self-sufficiency (including soil erosion, contamination of surface water, and a decrease in fishing stocks). These policies were unsustainable because of their costly and damaging nature.[19]

Since rice is a critical commodity for Indonesia, a lot of attention has been paid to agricultural policies and trade policies that affect the country's rice supplies and prices. After Suharto consolidated power in the 1960s, his New Order regime embarked on attempts to stimulate rice production and to improve the country's overall food security. This was a critical priority for Suharto; his predecessor, Sukarno, had failed to implement successful food policies and this was one factor contributing to high levels of social and political instability and volatility in the 1960s. Highly interventionist policies were pursued, with goals of bolstering food production, stabilizing prices, and, ultimately, regime stability. When there was an excess supply of rice, the state subsidized the farmers by buying the surplus. When there was excess demand for rice, the state subsidized imports of rice from foreign sources. This was a successful strategy on two fronts: (1) it provided price stability to farmers, which gave them confidence to make necessary investments to raise productivity; and (2) with increased productivity, prices decreased. However, like other interventionist policies, this approach became quite expensive in the longer run. Additionally, these policies created a bias toward rice farmers, which caused difficulties in diversifying Indonesia's other agricultural sectors.[20] Since 2004, there has been a ban on rice imports by private traders. Only Bulog is allowed to import rice.[21]

Over the past few years, Bulog has imported a significant amount of rice. This has been true even when domestic harvests have been quite good. Some of the reasons for importing foreign rice are to keep the prices low for consumers and to combat or counteract smuggling. When it comes to food security, particularly the issue of rice cultivation and trade, Indonesia has not made a concerted effort to change the way that trade and reserves are structured. This may in part stem from Indonesia's position of being mostly self-sufficient in rice production and, thus, it neither imports huge quantities of rice nor is it economically reliant on exporting rice to its neighbors. In 2010, Indonesia had a sufficient supply of rice from domestic production and allowed imports of only specific types of rice in the amount of 250,000 tons.[22] Then, in 2011, Indonesia bought a considerable amount of rice, 1.9 million tons. For the past few years Indonesia has imported rice in similar amounts; for example, in 2014 it imported 1.1 million tons.[23] One of the other problems that Indonesia faces (it is not the only country facing this issue) is that farmers are aging. The younger generation in rural areas is more likely to move to urban areas to seek a different lifestyle than that of a small farmer or peasant.

Indonesia is building a 480,000-hectare agricultural estate, the Merauke Integrated Food and Energy Estate, in West Papua. When it is finished it will be able to produce 2 million tons of rice, 2 million tons of corn, 167,000 tons of soybeans, 2.5 tons of sugar, and 937,000 tons of palm oil and provide grazing space for 64,000 cattle.[24]

Singapore

Singapore has traditionally been a passive consumer of food, importing virtually all manner of foodstuffs. It is now repositioning itself to create a more stable global food system. There are four parts to Singapore's strategy:

1. As a well-respected center of higher education, Singapore can focus resources on research and development in the areas of agriculture, aquaculture, and adaptation to climate change.
2. Singapore can reposition itself as a hub for agribusiness. The Singapore Economic Development Board is supporting the setup of operational headquarters and trading operations and research facilities for companies on the cutting edge of developing new crop varieties for the region.
3. Even densely populated Singapore sees potential in urbanized solutions to food security. Thus, it is experimenting in urban farming, rooftop farming, vertical farming, hydroponic farming, and the like.
4. Last is the goal of achieving more local production of three key food items: eggs, leafy vegetables, and fish. A $20 million food fund, launched in December 2009, is in place to provide incentives to explore new farming technologies and to increase Singaporean production capacity.[25]

Both government and private resources are being mobilized toward these goals, another example of the hybrid nature of transboundary problem solving. Singapore is also partnering with foreign companies abroad to explore other ways of ensuring access to food. Development is under way for the China-Singapore Jilin Food Zone. The first rice crop from the new food zone hit Singapore markets in December 2015, and pork is projected to be on the market in 2017.[26] Jilin province in China has fertile soil, water resources, and good growing conditions for a wide variety of crops. The full plan is for a 1,450-square-kilometer superfarm (more than twice the size of Singapore) that will be able to produce staples (e.g., corn, soybeans, and rice), grains (e.g., wheat, barley, oats, and rye), sugar beets, poultry, pork, beef, dairy products, fruits, vegetables, flowers, and roots and herbs.[27] Singaporean citizens and firms have also been buying land in Thailand for rice cultivation. Singapore is clearly looking for a multitude of different options for dealing with food supplies in the future.

The Philippines

The Philippines too is seeking to increase its own food security. Rice is an integral part of the history, culture, and, especially, the diet of the Philippines. There is a great deal of anxiety in the Philippines that the country is not self-sufficient in rice production. Some of the reasons for this include: urbanization, conversion of rice land to other uses, deteriorating irrigation systems, lack of access to credit for farmers, and poor government policies.[28] The Philippines is launching a site for production of "golden rice" (rice with vitamin A in it). Besides pioneering the use of this genetically modified strain of rice, it is moving ahead with the use of Bt (genetically modified) corn, of which it was an early adopter in Asia in 2002. Golden rice promises higher yields, and Bt corn is supposed to be hardier and higher yielding.[29] This initiative is in keeping with broader efforts to bolster agriculture. The government has initiated a campaign, Agri-Pinoy, to guide various services and government programs relating to: food security and self-sufficiency, sustainable agriculture and fisheries, natural resource management, and local development.

While these domestic policies may improve an individual country's food security posture, by themselves these policies may not be enough. The Philippines has struggled to raise citizens out of poverty, but poor policymaking and implementation as well as high levels of corruption have hampered economic growth. In addition, the country has been plagued by natural disasters and weak agricultural productivity. The

Philippines will need to improve its own efforts to alleviate food insecurity, and it will also need favorable regional and international trade conditions to ensure access to key food supplies such as rice. Only through cooperation between food exporters and food importers will longer-term prospects and mechanisms for dealing with food insecurity and periodic crises be improved.

Rice Exporters

Thailand

Thailand is one of the world's largest exporters of rice, particularly of high-quality fragrant rice (Jasmine rice). As part of Thailand's development strategy in the 1960s, rice farmers were paid artificially low prices for their rice by the government. Cheap rice was then available to urban dwellers, and smaller and less productive farmers were forced to sell their land and become tenant farmers or join the waves of Thais moving to urban areas. Over time, rice cultivation became a larger-scale operation conducted by bigger producers. Instead of being on the losing end of policies favoring industry, Thai farmers are now on the receiving end of political favoritism. Making good on an election promise, former Pheu Thai Party prime minister Yingluck Shinawatra reinstated a previous rice support program. The government paid farmers $490 (15,000 baht) per ton for paddy rice, which was well above market prices of around $295 (around 9,000 baht). This allowed the government to build up a stockpile of nearly 7 million tons, and it has improved prices for farmers.[30] However, it has hurt Thai rice exporters dramatically because Thai rice has become vastly more expensive and less competitive on world markets than rice from Vietnam, India, or Cambodia. Thai rice exports, in fact, were halved from only a few years ago due to the high prices; Thailand may now only export between 6 million and 7 million metric tonnes, the lowest levels exported since 1999–2000.[31] Domestic Thai politics has been chaotic and sometimes violent over the past five years. The Pheu Thai Party's victory in 2011 was viewed as a return to power of politicians who tended to favor rural interests over Bangkok elites. Yet even after the 2014 coup, the Thai government under Prime Minister Prayuth Chan-Ocha provided subsidies for rice farmers. The rice supports may have helped farmers in the short term but it has hurt exporters and, now with the two-year drought making rice cultivation more difficult, the government is actively encouraging rice farmers to diversify and start growing things such as limes and other fruit trees.[32]

Vietnam

Vietnam has been the second-largest exporter at an average of 6.1 million metric tonnes per year. Before the El Niño drought, Vietnamese exports reached close to 7.0 million metric tonnes. Unlike Thailand, Vietnam's exports are going in an upward direction.[33] The government of Vietnam plays a significant role in controlling the rice trade in Vietnam. The government ministry, the Vietnam Food Association, oversees rice cultivation and trade. The government is promoting efforts to raise the quality of Vietnam's rice, increase the crop yields, and promote exports.[34] Vietnam is also contemplating greater cooperation with Thailand on creating a rice cartel under the umbrella of a regional federation for rice among ASEAN rice-producing nations. The hope is that rice exporters will be able to command higher prices for their farmers if they work together as a group to set prices and control supply.[35] There are significant obstacles to this, however. Rice-producing countries vary in their approaches to cultivation, storage, and logistics capabilities. The partnership could mean sharing of storage facilities and higher prices for Vietnamese rice. This could weaken Vietnam's competitiveness, but could bring higher prices to its producers and the middlemen in the rice trade.

Cambodia

Cambodia is trying to break into the ranks of large exporting countries like Thailand and Vietnam. Cambodia's rice is much cheaper than Thailand's and Vietnam's, but much of the rice that Cambodia exports first goes to Vietnam for processing. Cambodia does not have enough good quality mills and, because of weak trade networks between producers, millers, and traders within Cambodia, it has been easier to send the rice to Vietnam for processing.[36] Cambodia is looking to join forces with Myanmar to cooperate on rice production and export. Cambodia would like to see the establishment of a rice cartel of Southeast Asian exporters, as it would help offset fluctuating international market prices. Cambodia is interested in bringing in Myanmar because Myanmar is in a relatively similar position to Cambodia. Since Cambodia and Myanmar are lower on the production chain, they have fewer opportunities for making money on the processing side of the rice trade and, thus, are even more vulnerable to market variability. In measures aimed at improving Cambodia's position in the rice trade, it plans to buy a rice-polishing machine from China and is also looking into building a rice storage facility in Kandal province.[37] Cambodia has reached out to China and the Philippines to sign trade contracts for rice with those two growing importers.[38] The interests of Thailand, Vietnam, and Cambodia should be relatively similar. All produce rice for export and all want to ensure open markets for this key crop. However, internal politics has trumped broader cooperation that could benefit all parties. Thailand's rice supports were enacted to benefit rural Thai rice farmers, but one of the effects has been a boost to Vietnam's and Cambodia's exports. Vietnam has benefited from Cambodia's reliance on it for processing, and so it too has conflicting interests on the benefits and liabilities from greater coordination. Creating a rice cartel would solve some of the problems with supply and pricing volatility. However, Thailand, Vietnam, and Cambodia have done little to get it off the ground. This demonstrates that in fact they may be more influenced in their behavior by domestic political considerations than by economic incentives. Coordinating rice production, processing, and trade among these three countries would require more extensive cooperation among private and state-owned firms and government planners, something which is hard to achieve.

Areas of Cooperation

While regional discussions about food security predate the rice crisis in 2007, it was partly in response to the sharp increase in international food prices in 2007–2008 that leaders in Southeast Asia pledged to embrace food security as a matter of permanent and high-priority policy and adopted the "Statement on Food Security in the ASEAN Region," which commits, among other actions, to the implementation of the ASEAN Integrated Food Security Framework and the Strategic Plan of Action on Food Security in the ASEAN Region (2009–2013). This was a five-year strategic plan to assure long-term food security in the region through increased cooperation and mutual help.

The Senior Officials Meeting (SOM) is the main ASEAN body that oversees ASEAN cooperation in food and agriculture, with the guidance of the ASEAN Ministers on Agriculture and Forestry (AMAF). Sectoral working groups, joint committees, boards, and groups of experts have been established to implement the respective cooperation sectors on food and the various subsectors of agriculture and forestry. The ASEAN Secretariat is charged with overall coordination of programs. ASEAN has implemented numerous cooperative projects relating to food, agriculture, and forestry sectors, which cover a wide spectrum of activities ranging from exchange of information, crop production, postharvest processing and handling, training and extension, and research and development as well as trade promotion in the areas of crops, livestock, fisheries, and forestry.

To respond to trade globalization, ASEAN cooperation in food, agriculture, and forestry is now more focused on the enhancement of food, agricultural, and forestry products' competitiveness in international markets while sustaining agricultural production. Harmonization of quality and standards, assurance of food safety, and standardization of trade certification are among the priorities being addressed. These measures build on the experience some member states have now gained by complying with existing international standards.

At ASEAN meetings in 2010, ministers welcomed and reaffirmed the ASEAN Plus Three Cooperation Strategy on Food, Agriculture and Forestry with the goal of ensuring long-term food security and improving the livelihoods of farmers in the ASEAN and Plus Three (APT) countries (China, Japan, and South Korea). The cooperation strategy provides a comprehensive framework to foster cooperation among the APT countries in the areas of strengthening food security, biomass energy development, sustainable forest management, climate change mitigation and adaptation, animal health and disease control, and cross-cutting issues (i.e., enhancement of capacity building and human resource development; strengthening of information and knowledge networking and exchange; enhancement of productivity, quality, and marketability of agriculture and agricultural products; and strengthening collaboration on research and development). The ministers agreed to formalize the ASEAN Plus Three Emergency Rice Reserve (APTERR) as a permanent scheme for meeting emergency requirements and achieving humanitarian purposes. The ministers urged all ASEAN Plus Three countries to sign the agreement and signal their support through early implementation.

By 2010, some progress had been made in creating the ASEAN Food Security Information System (AFSIS), including early warning systems and information on agricultural commodities. These were created to provide an early warning mechanism to signal information on food supplies hopefully before a crisis is full-blown.[39]

At the 2011 ASEAN meetings, food security was again on the agenda. There was a renewed sense of importance and urgency at the 2011 meetings. Commodity prices were again climbing, and heavy rains in Thailand were threatening to affect the rice harvests. Additionally, the events of the Arab Spring heightened government concerns over volatility of food prices because of the clear relationship between food security and political stability. The protests throughout the Middle East and North Africa clearly showed the world that the combination of high food costs, few economic opportunities for advancement, and authoritarian regimes that deny citizens meaningful participation and accountability, coupled with state capitalist regimes that reward regime insiders with economic benefits while most citizens are locked out of greater economic and political opportunities, is a recipe for psychological feelings of relative deprivation and ultimately protest movements.[40]

The November 2011 ASEAN summit took place against the backdrop of widespread floods in Thailand and questions about yields of rice from the region's largest exporter. The unusually heavy floods may have provided the needed push to jump-start efforts to address issues of food scarcity. The ASEAN-related meetings had a full agenda of issues to address: negotiations on conduct in the South China Sea, political changes in Myanmar, and the ongoing dispute between Cambodia and Thailand as well as the usual discussions about reducing trade barriers and community-building efforts. With the intense rains, food security moved up on the list of priorities to address.

On October 12, 2011, as ASEAN chair for the year, Indonesia issued a special statement concerning the floods: "Jakarta expressed concern over the rising waters deluging some member states. But it also 'emphasized the importance of strong cooperation and coordination' among states and the 'need for full implementation' of two key mechanisms to deal with food emergencies."[41] The two mechanisms discussed were: the ASEAN Agreement on Disaster Management and Emergency Response (AADMER) and the ASEAN Coordinating Centre for Humanitarian Assistance on Disaster Management (AHA Centre). While these institutional structures exist, they have not yet been tested to see if they are ready to play their designated roles in addressing challenges to human security.

Also at the fall 2011 meeting, the ten members of ASEAN signed an emergency rice reserve agreement with their three Northeast Asian dialogue partners (China, Japan, and Korea), the ASEAN Plus Three. Based on earlier cooperation, the APT signaled their support for the APTERR. Under the current plan, the thirteen signatory states will stock up 787,000 tons of rice to be used in the event of a sudden instability in rice supplies. This reflects clear nervousness among regional leaders over environmental conditions and the close connection between governments' ability to provide public goods such as water and food supplies, costs, and domestic stability.[42] Indonesia's president Susilo Bambung Yudhoyono consistently called on ASEAN to consider using APTERR not just for responding to natural disasters, but also in the event of economic upheaval. This clearly reflects political considerations in light of the experience of shocks to the rice market as happened in 2008, and the uprisings in the Arab world, partly driven by economic malaise and rising food costs throughout the world. It should also be noted that having a rice reserve could be even more significant in a psychological way than as a practical solution to food supplies. It was the announcement of the okay to use

WTO rice stockpiles in Japan in 2008 that ended the hoarding that caused the price spike. The WTO stockpiles were never actually used. APTERR could play a similar role in providing psychological stability to the rice markets and, thus, it could prevent dramatic price fluctuations just by its very existence.

Signaling that food security is also a big issue outside of Southeast Asia, the East Asia Summit, involving the APT and Australia, New Zealand, India, the United States, and Russia, has also taken up the concern. Because the United States is not a member of ASEAN, nor of APT, it is hard to see how the United States could play a more significant role within the context of these institutions in addressing regional food cooperation.

Prospects for Greater Cooperation

Timmer has pointed out a contradiction in the area of food security policy: international donors discourage countries from stabilizing domestic food prices in the face of a crisis. This sort of intervention in the market is frowned on by liberal economists pushing for market freedoms and less government distortion. Instead, donors encourage a policy of social safety nets, proscribing that governments should have these in place to help the poor. Thus, when a crisis hits, the poor will be able to cope with price fluctuations. While these policies may make long-term sense, they are not useful for quick reactions when a food crisis looms. Longer-term planning and the creation and maintenance of safety nets for the poor are certainly desirable, but they are costly and take years to put into place. So, more active measures are needed to prevent food price spikes, both domestically and internationally. Timmer recommends strategically held stockpiles that could be used to offset price jumps in certain situations. Reserves could serve two different, but useful, purposes. First, the stockpiles could be used to offset periods of scarcity in limited situations. More importantly to politicians perhaps, the reserve could act as a stabilizing force in the event of instability in the supply or price of a particular crop. Such reserves could increase market stability for basic commodity crops such as rice, maize, and soybeans.[43]

The discussions at the November 2011 ASEAN summit were promising. APTERR aims to do exactly what Timmer is calling for. However, if the past is any indication, it may be harder than it seems to get agreement on more structured cooperation; for example, on the logistics of the strategic stockpiles for rice reserves, greater transparency in the rice markets, or greater liberalization of export controls over rice. WTO discussions and rules almost all focus on import barriers rather than export controls. However, there is widespread agreement that price instability occurs because of export manipulation as well. It is difficult to imagine countries being willing to give up policy control on the export side, especially when it is the export of a stable food crop that is up for discussion. States tend to view food supplies and control over food supplies as a question of national security. So, it is hard to imagine large food exporters wanting to be constrained in how they control the distribution of such resources. Of course, there is also the basic problem that the interests of food exporters and food importers are dramatically different. In Southeast Asia, Vietnam and Thailand are big exporters of rice, and the Philippines and Malaysia are significant importers of rice. Indonesia is in an interesting position; from year to year, it varies whether it imports significant supplies of rice, or whether it is actually exporting some tonnage of rice. Naturally, the interests of countries on both sides of the rice trade vary tremendously.

In theory, both importing and exporting countries would both benefit from greater transparency and greater liberalization of the rice trade. There could be large payoffs from greater efficiency and better allocation of resources: greater stability of prices and supplies, less of a tendency for actors to hoard stocks, and a reduction in the likelihood of price panics. So, why don't countries liberalize trade in rice and increase their cooperative efforts? Timmer views this cooperation challenge as a Prisoners' Dilemma. All countries would be better off if everyone cooperated. However, if a country defects while other countries cooperate, then the defecting country will come out ahead. This then generally produces the understandably rational decision of each individual country to forgo cooperation in favor of beggar-thy-neighbor policies (even though cooperation by all would produce the best outcome). Timmer explains that the ways to resolve this dilemma are: (1) agreement on a formal mechanism for cooperation (through ASEAN or a related organization) with binding commitments such as General Agreement on Tariffs and Trade (GATT)–WTO trade agreements that have governed other areas of trade, but in which rice is currently exempt; or, alternativcly, (2) gradual learning by all actors to produce greater cooperative outcomes in a Prisoners' Dilemma. That is, through repeated rice price crises, countries may come to realize that cooperation is the best way forward. This, of course, is a painful way to get to greater cooperation.[44]

Another problem relates to prospects for greater cooperation on rice in Southeast Asia: it is the nature of the market itself. The rice market is relatively small (in comparison with other commodity markets), and it is highly segmented and "imperfect" as economists would say. Most import and export deals on rice are made behind closed doors by governments without much accountability.[45] Weak cooperation on the rice trade exemplifies a high degree of both state control and fragmentation.[46] The creation of APTERR, the strategic rice reserve, and AFSIS, the monitoring system for food supplies, demonstrates the potential for synergistic fragmentation. In both, the processes, responsibilities, and components of the larger plan are delegated to various states and agencies. Making these regimes work requires little actual cooperation; instead, each agency and actor has a distinct role to play independent of the others. Neither states, nor private actors such as farmers, agribusiness interests, or exporting firms, are being asked to give up power or sovereign control to achieve the higher good of cooperating. While these mechanisms have been created, they have yet to be put to the test to see if they can function as designed and keep a crisis from escalating.

Conclusion

Like other issues discussed in this book, regional attempts at cooperation on food security demonstrate the difficulties of Southeast Asian states to (1) improve governance systems; (2) implement policies that expand the provision of public goods over more narrow private interests; and (3) work cooperatively within regional and international organizations and with other states to solve common problems. Failure to overcome problems in these three areas may heighten risks of food insecurity for millions of citizens in the region. It will take significant leadership to envision greater cooperation on food security in Southeast Asia. Since Indonesia is the largest power in Southeast Asia, and it is showing renewed interest in acting like a regional great power,[47] it would behoove Indonesia to push for longer-term goals of policies to engage in better integration of the Asian rice market with more open trade policies and much greater transparency. As countries in the region continue to see strong economic growth rates and increasing levels of prosperity, this may get easier. As populations get richer, they will be better able to afford different diet options and demand for rice may shrink. As demand drops, states may have greater flexibility in how they view their control and access to rice supplies. Leadership on cooperative issues is vital and as hegemonic stability theory demonstrates, stability is enhanced or even provided when a superpower is willing to act to promote the greater collective good. One way of improving the chances of seeing cooperative behavior is to lead by example or, even more effectively, to incentivize others to go along with cooperation by making it worth their while to do so. In the case of the rice trade, leadership ideally should come from a country that could push for greater openness in the rice trade and for more robust legal frameworks for export rules. Since Indonesia is not a significant rice exporter, it is unlikely to be able to lead in this way. No other country in the region is acting in this way either.

Indonesia could continue to push for larger and better coordinated rice reserves in member states of the ASEAN Plus Three. Realistically, reserves would not be of a size significant enough to prevent a crisis, but it could increase confidence and thus act as a psychological buffer to a crisis, much as releasing the WTO rice stocks in Japan did in 2008. This kind of limited stockpiling of rice under the control of the regional organization has been under discussion for a number of years with little firm progress made. Timmer suggests that having Australia serve as the coordinator and holder of a reserve supply of rice for the Southeast Asian region might also work. Australia is a relatively small market for rice and can serve as a trusted neutral player in this role.[48] Again, however, Indonesia is in an unlikely position to be able to act as the driver of this initiative. Indonesia and Australia do not have close relations with each other and again, since Indonesia is not a major rice exporter, they would not have the same weight to prod their neighbors into working on this.

Asia to a large extent has been able to manage keeping domestic food prices low, but this has had a negative externality for world markets. The 2007–2008 rice panic illustrates important lessons regarding food security. The reactions to India's decision to curtail rice exports reflect how price expectations for basic food grains are formed by farmers, traders, governments, and consumers. The crisis shows how critical it is to know what policy actions can be taken to stabilize food prices and keep consumers more food secure. To do this, Asia needs to find more efficient ways to stabilize domestic food prices, particularly for rice, without harmful spillover effects on domestic producers, on world markets, or on the environment.

Rice is increasingly a food consumed by the poor. As people climb out of poverty, their diets become less dependent on rice. To deal with food security based on adequate availability of low-cost rice is to deal with only part of the problem. Food security is very much a part of larger questions. One of those questions relates to how countries deal with the wider problem of poverty. Policymakers must implement long-term measures to stimulate pro-poor economic growth and sustain that growth for decades to have a measurable impact on poverty rates. At the same time, countries need to have short-term measures to address everyday issues of food security as well as policies in place to deal with possible crisis. These are the challenges that policymakers in much of Southeast Asia face and, at their core, these are questions of human security.

Food security is also clearly linked to the larger challenges posed by climate change. Climate change will have profound effects on living conditions and human security and will impact many diverse issues: water resources, agriculture, forestry, health, migration patterns, energy systems, and, arguably, the larger economy itself. The Center for Strategic and International Studies stated in November 2007 that climate change "could destabilize virtually every aspect of modern life."[49] Food insecurity in this context is just one of a myriad of issues that will require long-term planning and regional cooperation. One of the reasons that human security issues are often hard to tackle, as the issues in this book so clearly illustrate, is that we live in a world that is still defined by states and national security interests. Addressing human security requires a shift in focus from being concerned about state-centered approaches to security to focusing on people-centered and transboundary approaches to security. This will require a rethinking of notions of security, priorities, and actors as well as vastly more cooperation across state borders. This is incredibly difficult, but not impossible. Governments do see a link between human security and state security; if for no other reason than most governments in the region feel some vulnerability to the potential of domestic instability. Since food insecurity, like the water security and migration issues discussed in Chapters 4 and 6, has the potential to drive people into the streets, governments are at least somewhat mindful of being able to provide stable markets for food. Thus, while governments in the region are not moving dramatically or quickly to improve cooperation and domestic governance on this problem, there are efforts under way.

In 2009, Indonesia and Malaysia initiated a series of bilateral discussions about cooperating on food security. The two countries stated their intention to jointly develop investment programs in agriculture. Indonesia has vast agricultural assets, and Malaysia is more technologically advanced. The goal would be that Malaysia would invest capital and technology in Indonesia's agricultural sector for mutual benefit. This would enhance regional cooperation and serve as a model for more multilateral efforts within ASEAN, and it would improve food security and help prevent food crises.

Successfully solving cooperation problems often requires strong leadership. Since the largest rice exporters in Southeast Asia are not the region's most powerful actors, it is difficult to predict greater, more successful, or more robust cooperation in the near term. Indonesia should be in this position as regional leader but, as a frequent importer of rice, it is harder to view how it could take the lead. Price instability over the past few years, and prospects of more fluctuations in the future, may be a catalyst for greater action. As the region's largest exporter, Thailand could be the leader in promoting cooperation on the rice trade. However, domestic politics in Thailand are highly contentious,, and the decision to reinstate rice subsidies for Thai farmers and the blowback from exporters and business groups show how policies there (like elsewhere) are driven by internal dynamics, not necessarily by regional concerns. What we saw in the ASEAN meetings in the fall of 2011 may be the beginning of an attempt to put regional cooperation more squarely on countries' agendas. But it will take regional leadership for more to be accomplished. The attempts by Thailand, Vietnam, and Cambodia (and Myanmar) to cooperate on exports and to move toward a more cartel-like organization is an example of cooperation and an example of hybridity (in that the actors are both government procurement agencies and some private producers and export agencies). But it is not clear how this cooperation would benefit the region more generally.

While countries are choosing beggar-thy-neighbor policies, there are also very real attempts to build institutions and to forge cooperation on issues of common concern. These self-interested policies are more a product of domestic politics than decisions based on security or sovereignty fears. Thailand's decision to support rice farmers and India's decision in 2007 not to export rice are products of domestic political concerns or interests, not security dilemmas per se. If states do not in fact view access to food as a security threat, they may be more willing in the future to build more robust cooperative mechanisms.

More work needs to be done to increase transparency in rice markets to build institutional cooperation; to provide information on rice supplies, prices, and so forth; and to balance the needs of producers and consumers. There must be greater regional leadership from exporters like Thailand, Vietnam, and Cambodia. This is an important piece of the puzzle of trying foster greater cooperation between importers and exporters. The cooperative efforts that have happened through the APT are an important step, but those efforts have not yet played a significant role in reshaping domestic agricultural policies.

Notes

1. World Health Organization, "Food Security and Famine Glossary of Terms," 2012, www.who.int/trade/glossary.

2. Ibid.; see also Colin D. Butler, "Food Security in the Asia-Pacific: Climate Change, Phosphorus, Ozone and Other Environmental Challenges," *Asia Pacific Journal of Clinical Nutrition* 18, no. 4 (2009): 590–597.

3. S. Rajaratnam School of International Studies (RSIS) Centre for Non-Traditional Security (NTS) Studies, *International Conference on Asian Food Security*

2011—Feeding Asia in the 21st Century: Building Urban-Rural Alliances (10–12 August 2011) (Singapore: RSIS NTS), 7.

4. Ibid., 8.

5. Organisation for Economic Co-operation and Development (OECD), "Share of ODA for Agriculture (percent)" (Paris: OECD Creditor Reporting System, 2009).

6. Le Danh Tyuen, "Food and Health Security in Southeast Asia," *Asia Pacific Journal of Clinical Nutrition* 18, no. 4 (2009): 493–497.

7. For extensive information on swidden, see Jefferson Fox, Yayoi Fukita, Dimbab Ngidang, Nancy Peluso, Lesley Potter, Niken Sakuntaladewi, Janet Sturgeon, and David Thomas, "Policies, Political–Economy, and Swidden in Southeast Asia," *Human Ecology* 37, no. 3 (2009): 305–322.

8. "International: Biofuels Spark Food-Versus-Fuel Debate," *Oxford Analytical Daily Brief,* December 7, 2007.

9. Ibid.

10. Ibid.

11. Robert Siegel, "How Fear Drove World Rice Markets Insane," National Public Radio (NPR), November 2, 2011.

12. Tyuen, "Food and Health Security in Southeast Asia."

13. Siegel, "Fear Drove World Rice Markets Insane."

14. Ibid.

15. Daniel Workman, "Rice Import Dependent Countries Nigeria, Saudi Arabia and Philippines Among Biggest Rice Buyers" (Los Baños, Philippines: International Rice Research Institute, April 16, 2008).

16. David C. Dawe, "The Philippines Imports Rice Because It Is an Island Nation," in *Why Does the Philippines Import Rice? Meeting the Challenge of Trade Liberalization,* ed. David C. Dawe, Piedad F. Moya, and Cheryll B. Casiwan (Los Baños, Philippines: International Rice Research Institute and the Philippine Rice Research Institute, 2006); "USDA Post Estimates for Malaysia's Rice Production, Imports," *Oryza,* March 3, 2016.

17. Butler, "Food Security in the Asian Pacific."

18. Hanim Adnan, "Food Security Policy May Be Extended—Business News," *The Star Online,* April 10, 2011, www.thestar.com.my/business/business-news/2011/04/09/food-security-policy-may-be-extended/.

19. Edward B. Barbier, "Cash Crops, Food Crops, and Sustainability: The Case of Indonesia," *World Development* 17, no. 6 (1989): 879–895.

20. C. Peter Timmer, "Food Security in Indonesia: Current Challenges and the Long-Run Outlook," Working Paper No. 48 (Washington, DC: Center for Global Development, November 2004).

21. "Indonesian Rice Production and Trade Policy," AsiaOne, March 3, 2011.

22. US Department of Agriculture, "USDA Foreign Agricultural Service GAIN Report: Indonesia Rice and Corn Update" (Washington, DC: USDA, April 23, 2010).

23. Data retrieved from the UN Food and Agriculture Organization, "Food and Agricultural Data," updated information from August 14, 2017, http://faostat3.fao.org/faostat-gateway/go/to/download/T/TP/E.

24. "Indonesia Cannot Afford to Ignore Food Security," *Jakarta Globe,* January 8, 2011.

25. Yang Razali Kassim, "ASEAN's Role in Asian Food Security," *Business Times,* September 14, 2011.

26. Grace So, "Rice Pork and More to Come From Jilin Food Zone," *Business Times,* March 3, 2016.

27. Sonia Kolesnikov-Jessup, " Singapore Looks to China for Food Security," *New York Times*, September 27, 2010.

28. Dawe, "The Philippines Imports Rice Because It Is an Island Nation."

29. Melody Aguiba, "Philippines to Play a Critical Role in Global Food Security with 2012 Golden Rice Launch," *Jakarta Globe,* March 19, 2011.

30. Ross Korves, "Thailand's Domestic Rice Policies Upset Trade Patterns," Global Farmer Network, February 9, 2012; "Thai Rice Exports Slump on Intervention: Vietnamese Exports Surge," *Business Recorder,* April 22, 2012.

31. Korves, "Thailand's Domestic Rice Policies Upset Trade Patterns," 1.

32. "Low Rain Forces Thai Rice Farmers to Consider Growing Other Crops Instead," *Japan Times,* April 8, 2016.

33. "Rice Shipments from Vietnam Reach 2.87 MMT to June 14th," Bloomberg News, 2012.

34. For more on government efforts, see Government of Vietnam, "Vietnam Food Association: Introduction." Vietfood rice information last updated 2011, www.vietfood.org.vn/en/default.aspx?c=111.

35. "Vietnam to Cooperate with Thailand on ASEAN Rice Cartel," *Angimex,* June 26, 2012, www.angimex.com.vn/en/2012/06/vietnam-to-cooperate-with-thailand-on-asean-rice-cartel/.

36. "Rice Cartel Plan Resurfaces," Radio Free Asia, March 21, 2012.

37. May Kunmakara, "Phnom Penh Autonomous Port to Push Rice Exports," *Phnom Penh Post,* June 8, 2012.

38. "Philippines, Cambodia to Sign Rice Trade Deal," *Manila Times,* May 3, 2012.

39. "Second Protocol to Amend the Agreement on the ASEAN Food Security Reserve," Joint Press Statement of the Tenth Meeting of the ASEAN Ministers on Agriculture and Forestry and the Ministers of Agriculture of the People's Republic of China, Japan and the Republic of Korea (10th AMAF Plus Three), Phnom Penh, October 24, 2010.

40. Ted Gurr, *Why Men Rebel* (Princeton: Princeton University Press, 1970).

41. Yang Razali Kassim, "The 19th ASEAN Summit: Tackling Floods, Food and Stability—Analysis," *Eurasia Review,* November 17, 2011.

42. Ibid.

43. C. Peter Timmer and Thom Jayne, "Managing Food Price Volatility: Approaches at the Global, National and Household Levels," paper presented at the "Center on Food Security and the Environment: Stanford University Symposium Series on Global Food Policy and Food Security in the 21st Century" (Stanford: Stanford University, May 26, 2011).

44. Ibid., 14.

45. Ibid., 11.

46. Frank Biermann, Philipp Pattberg, Harro van Asselt, and Fariborz Zelli,"The Fragmentation of Global Governance Architectures: A Framework for Analysis," *Global Environmental Politics* 9, no. 4 (2009): 14–40.

47. During Indonesia's time chairing ASEAN in 2011, it pushed hard to strengthen the capacities and role of the ASEAN Intergovernmental Commission on Human Rights (AICHR), pursued the goal of establishing a network of peacekeeping centers in ASEAN member states, and laid the groundwork for the Southeast Asia Nuclear Weapons Free Zone (SEANWFZ) Treaty to be signed by the United States, United Kingdom, Russia, France, and China. These are part of Indonesia's growing desire to be seen as a leader on issues of human rights, peacemaking, and stability among great powers. Lina Alexandra, "Indonesia Emerging," *The Diplomat,* August 11, 2011.

48. Timmer, "Managing Price Volatility," 23–24. There are significant criticisms of this plan. While Australia might be neutral in terms of the rice trade, Australia's foreign relations with countries in Southeast Asia is more generally marked by both

cooperation and conflict. So, it is not really a foregone conclusion (as C. Peter Timmer might have thought) that countries in the region would see Australia as a good steward for a regional rice reserve.

49. Jürgen Scheffron and Antonella Battaglini, "Climate and Conflicts: The Security Risks of Global Warming," *Regional Environmental Change* 11, no. 1 (2011): S27–S39.

4

Water Security: Droughts, Deluges, and Dams

Water is the most basic of human needs. Like climate change, migration, and food security, water security is a problem that affects countries in Southeast Asia in different ways and with differing levels of severity. While there are domestic, regional, and international organizations involved with helping countries manage their water resources, there are still significant problems to be addressed. Perhaps surprisingly, Asia is the most water-stressed continent in the world. Asia has 47 percent of the world's average of freshwater per person, but 65 percent of the world's population.[1] As discussed in Chapter 3 on food security, one of the ironies at work is that, as the region's population becomes more prosperous, water supplies come under even greater strain. Intensified industrialization, mechanization of agriculture, shifts in dietary habits toward more animal proteins and dairy and away from vegetables and legumes, and growing urbanization all increase per capita water use. Combining this with the effects and pressures of climate change—which have resulted in larger storms, more severe weather conditions such as the 2016 El Niño, too much water in some places and not enough water in other places—has meant that water resources are increasingly important to manage efficiently, sustainably, and cooperatively. Yet this management is not happening fast enough in Asia.

In this chapter we look at the problem of water security in general, and then focus particularly on the question of how the six countries located along the Mekong River (China, Myanmar, Cambodia, Laos, Vietnam, and Thailand) are able to cooperate on this vital and shared resource. Water security is a broad topic and many of the issues relating to access to clean water for drinking and sanitation, agriculture, and protection from flooding and drought are solely domestic issues. However, there are several specific water issues that are transboundary in nature and are crucial for citizens in Southeast Asia: access and rights to fishing grounds and shipping lanes in contested waterways are a constant source of tension in the region; and the shared use of the Mekong River among riparian states is becoming more problematic. In this chapter, we focus on what is happening on the Mekong River. Shared use of this rich resource is vital for China, Myanmar, Vietnam, Cambodia, and Laos. It is clearly a problem that transcends state boarders: when damming the river changes the flow of silt downstream and to tributary rivers and lakes (fed by these rivers), and when 60 percent of Cambodians' protein needs come from these waterways,[2] then clearly what happens outside of Cambodia's borders (just to use this one example) can have devastating consequences within the country. Security experts are increasingly factoring in water insecurity as a potential trigger for military conflict among states, so this is an issue that has the clear potential to rise from being a nontraditional or human security threat to becoming a military security threat. Thus, water usage and rights need to be looked at as a human security issue as well as a national security issue. If dams up stream in China or Laos cause dramatic changes in the river flow and composition, is this an infringement on lower states like Vietnam's or Cambodia's territorial integrity? It certainly could be interpreted this way, particularly in the direst set of circumstances. In Southeast Asia, China is increasingly in the driver's seat. It is at the forefront of dam building, in China and in Laos, on the Mekong, and over time it will have greater and greater control over the river's resources.

Since countries along the Mekong need to cooperate to enjoy the benefits of the river's rich resources, it is imperative to understand how well or how poorly they are doing so. Access to the river has been treated like a limitless public good. Governments along the Mekong have increasingly viewed and used the river as a private resource that they alone can use as they like. Used in this way, the river will not be an infinite resource and, thus, countries' usage needs to be viewed as rivalrous. If all riparian countries are to be able to protect their citizens' access and use of the water, then greater long-term planning, cooperation, and coordination is necessary. In this chapter we show that there have been some attempts to share information and cooperate, but that there has been little committed interest from China (the country of origin for the river) on creating legal and binding agreements to cooperate on use of the river. In the case of water security on the Mekong, China is the undisputed hegemon. However, it has little apparent interest in viewing the river as a regional public good to be shared for the benefit of all. Instead, China seems to view the river as its resource to share as it sees fit. We demonstrate that countries have made attempts to cooperate with each other over the use and legal framework (governance) of the Mekong, but there is little to show for these efforts, and the lack of constructive commitment from China make prospects for improvement in the future doubtful.

Additionally, like other problems discussed in this book, countries' domestic policies often are not up to the task of solving basic human development needs and providing public goods.[3] This is because of center-periphery or fragmented governance problems, corruption, weak state capacity, or a combination of all three. States in Southeast Asia have not done a particularly good job in addressing water management issues. Instead, countries have relied on increasing their usage of existing water sources (in irrigation for agriculture, in electricity generation, and for increasing urbanization and industrial uses) rather than focusing efforts on sustainability and management of existing resources.[4] This challenge is complicated by the recent dam building, involvement in Southeast Asian affairs, and increased power of China. There is little room for optimism that China will want to play anything other than the role of (self-interested) hegemon in determining how, how much, when, and why shared water in the Mekong will be used.

Water Security and Insecurity

Water security is defined by the following five benchmarks. Citizens are water secure when these conditions are met:

1. Household water and sanitation needs are met in all communities;
2. Water supplies are sufficient to support productive economies in agriculture, industry, and energy;
3. There is ample water to develop vibrant livable cities and towns;
4. Healthy rivers and ecosystems are restored; and
5. Communities are resilient and able to adapt to change.[5]

In Southeast Asia, no country other than Singapore has achieved these goals. Southeast Asians are water insecure! Despite the perception that Southeast Asia is lush, rainy, and dotted with rain forests and jungles, it is in reality one of the driest subregions on the continent in per capita terms. It faces a number of water-related challenges: urban flooding, the need for greater water infrastructure to provide citizens with clean drinking water and water treatment capacity (the need for sanitation infrastructure), the depletion of groundwater and aquifers, and the shared used of common waterways (oceans and rivers). In all of Asia, about 1.2 billion people do not have proper sanitation (clean drinking water and waste treatment facilities).[6] Increasing urbanization is occurring faster than governments can provide sanitation. City dwellers are also competing with farmers to consume increasingly scarce water.

> It takes about 1,000 tons of water to produce one ton of rice, Asia's main food staple. According to the UN's Food and Agricultural Organization (FAO), farmers will need an additional 19% of water by 2050 to meet the demands for food, much of it in regions already experiencing water scarcity. The FAO estimates that food output must rise 70 percent by 2050 to feed a world population expected to grow from 7 to 9.3 billion. This demand is foreboding in light of the projected decrease in water availability in the region due to climate change.[7]

This problem is exacerbated by changes in diet; as people become more economically prosperous, traditional plant-based diets often give way to diets based more on animal protein. This is enormously problematic as it takes more acreage, more water, more fertilizer, and greater use of pesticides to raise crops to feed chickens, pigs, and cows than to raise crops used directly as human food. As is, agriculture already consumes 71 percent of global freshwater use.[8] In Asia, this percentage is even higher; 84 percent of all water withdrawal is for agriculture.[9] Changing diets and population growth are projected to increase demands for food. The Food and Agriculture Organization and the International Water Management Institute (IWMI) fear that it will be impossible to meet these demands given the already stressed water sources.[10] Sierra Ewert notes that, for Asia to meet the rising demands of increased population and greater affluence, countries will need to do the following:

1. Value water as a precious resource while ensuring that all citizens have access to enough water for basic daily needs;
2. Improve regulation of water withdrawal to prevent water waste;
3. Update irrigation systems, many of which are outdated and inefficient;
4. Tap into the greater potential of water recycling and reuse;
5. Focus on rainwater management; and
6. Educate citizens on water scarcity, responsible water use, and water conservation.[11]

One of the water issues that gets the most attention in Southeast Asia is repeated flooding in some of the region's most populous cities. Flooding is a regular and repeated occurrence in Bangkok and Jakarta. From the summer of 2011 through January 2012, Bangkok and the surrounding regions suffered one of the worst periods of flooding in the past 100 years; 567 people died in Thailand, 248 in Cambodia, 85 in Vietnam, 30 in Laos, and 102 in the Philippines.[12] Estimates are that 2.4 million people in 28 of 76 provinces in Thailand were impacted, and that economic losses may have totaled $45 billion.[13] In the wake of the extensive damage, the Thai government has spent millions of dollars on dredging canals and enhancing dikes and floodwalls around urban and industrial centers. Jakarta faces chronic flooding as well. Major water inundation occurred in 1997, 2002, and again in 2007. In both Bangkok and Jakarta, rapidly exploding urbanization over the past fifty years has stressed water usage and disposal. There is a huge need to increase drainage capacity; the canals built throughout Southeast Asia were made for irrigation needs 200 years ago, not to control flooding in the megacities that exist today. The 2007 flood in Jakarta was also one of the city's worst; 70 percent of the city was deluged, 57 people were killed and 450,000 (of a total population of 17 million) people were displaced from their homes.[14]

In addition to the immediate loss of life from flooding, and the economic harm it causes, flooding is linked to the spread of infectious diseases. The risk of communicable illnesses, particularly fecal-oral diseases, increases dramatically in the wake of flooding. These diseases spread when bacteria from untreated fecal matter gets into water used for bathing, drinking, or preparing food. When areas are hit by flooding, waste management systems are overtaxed and knocked out of order; thus, sanitary waste removal and provision of safe drinking water often gets compromised in flood situations. These diseases include typhoid fever, paratyphoid, poliomyelitis, and hepatitis A and hepatitis E. Increases in mosquito-borne and rodent-borne illnesses are also likely.[15] Cities such as Jakarta, Bangkok, and Manila (among others) have pumps (on which they are spending significant sums of money to upgrade and improve capacity) that are used to move the water out of city streets, canals, and drainage systems to rivers and the sea to minimize flooding in urban areas. Flooding is a clear threat to human security and well-being for the rapidly urbanizing population in the region.

A lot of the water problems discussed above are problems in need of better governance, better policymaking, and better allocation of resources. Many countries will share best practices with each other, and there are international experts advising and partially funding efforts at addressing these problems. But for the most part, the challenges with water diversion and drainage, sanitation, increasing salinity, and depletion of groundwater sources are, and should be, addressed as domestic problems. This is emphatically not the case when we look at the shared use of rivers and oceans across Southeast Asia. One of the best examples of the need for much greater interstate cooperation and the need to create legal frameworks for governance is the effort to address the shared use of water in the Mekong River and its many tributaries and offshoots.

Mekong Cooperation and Competition: The Formation and Functioning of Regional Organizations

Countries along the Mekong are heavily dependent on the freshwater from the river and its tributaries, and over 70 million people live along the Mekong River Basin.[16] "People and environment in 97% of the land area of Lao PDR, 86% of Cambodia, and large areas of Thailand and Vietnam, meet their water needs from the Mekong River Basin (MRB)."[17]

A river basin organization can play a vital role in helping monitor usage of a river's resources, its health and environmental status, and countries' compliance with authorized uses as well as in investigating irregularities. Such an organization can also serve as a licensing authority to administer water use rights and as the arbiter and enforcer when rules and use regulations are violated. It can function as a tool for planning and coordination of the development and operation of transportation networks, oversee and structure development projects such as hydroelectric plants, coordinate economic planning for the river, and plan for a fair allocation of water supplies and benefits accrued from activities just listed.[18] The organizations discussed in this chapter relate to the larger question of water governance:

> Water governance refers to the way in which power is distributed, organized, shared, and negotiated in society, the interactions of stakeholders and decision-making processes involved in how water resources are to be developed and utilized, and benefits and involuntary risks distributed. This includes political, administrative, and policy structures, both at the formal and informal levels that are embedded in and influence decision-making and management practices.[19]

Cooperation and efforts to codify use of this shared resource go back to the 1957 Coordination Committee created by the UN Economic Commission for Asia and the Far East (UNECAFE). This committee was formed to facilitate cooperation and to explore joint management in development of the river for hydropower production, expansion of irrigated land, and flood reduction among the lower riparians: Thailand, Vietnam, Cambodia, and Laos. All recovering from war and all (except Thailand) newly independent after the end of colonialism, these countries sought to create a cooperative framework for use of their shared river. China and Myanmar were not involved in the organization but at the time neither had development plans for the upper Mekong; both countries were poor and had low levels of industrialization and no large-scale agricultural infrastructure projects planned for the river (and at the time the region was not facing water shortages or stresses).[20] It is interesting to note, that these early plans for cooperation were *not* set in motion because of some specific conflict over the river (or other flash point), or because an upstream nation was depriving downstream countries of being able to use the river for their own needs.

Early discussions were forward-looking and hopeful that a framework of cooperation could be created before discord and conflict might arise. The creation of these kinds of institutions reflects the development of a global norm to codify and create standards for cooperation. The Coordination Committee of the four lower Mekong countries convened in Bangkok as a Preparatory Commission in the fall of 1957 and endorsed a statute establishing the Committee for Coordination of Investigations of the Lower Mekong (Mekong Committee).[21] The organization was funded in part by the UN and by a donation from France. Technical assistance was given by the US Army Corps of Engineers. The primary outcome of the early years of cooperation was to plan for future hydrographic studies of the river. Then, networks of hydrologic and metrological stations were set up (which continue to function) along the river to map, survey, and level portions of the river. Early efforts bore fruit as navigation was improved along the river.[22] Early cooperation among these four countries did (as functional theory predicts) improve cooperation and trust among countries that viewed each other with greater hostility on a range of more political issues. In the 1970s, work of the Mekong Committee slowed due to political upheaval in the region and lack of finances. However, in 1975 (despite ongoing conflicts and tensions) the lower Mekong countries set out to modify the committee's objectives and principles for development; they laid out a Joint Declaration on Principles and were able to agree on a precise definition of "reasonable and equitable use" of the water resources. Funding for the committee's work came from UNDP, the United States, and the Ford Foundation, among other sources.[23] Over time, hundreds of surveys and studies came out of the Mekong Committee's creation.[24] By the 1970s to 1980s, the committee's research was geared disproportionately toward hydroelectric and irrigation uses for the river. The committee's early success demonstrates that technical cooperation is possible. Countries viewed the committee's work as benefiting everyone without asking any country to sacrifice their interests. This provides us with an example of success in providing modest public goods;

all countries benefited from the information gleaned from surveys and studies and, thus, they supported the committee's goals.

In 1992, the Asian Development Bank (ADB) initiated the creation of the Greater Mekong Subregion Economic Cooperation Program (GMS). The members are ASEAN 5 (Cambodia, Laos, Myanmar, Thailand, and Vietnam), plus China. The GMS, because its membership includes China and because of support from the ADB (for funding), could be a productive organization for addressing common economic concerns. Its focus is on strengthening economic collaboration among member states and on facilitating economic and transportation infrastructure (e.g., energy grids, power interconnections, regional trade) to better integrate neighboring countries.[25] It seems that participating countries view the GMS as useful for mobilizing technical and financial resources for these economic goals, not as an organization that can tackle the wider range of mutual interests and potential conflicts over use of the river's resources.

A 1995 Mekong agreement established the Mekong River Commission; a successor to the Mekong Committee (made possible because of the Paris Peace Agreement ending the long-running conflict in Cambodia). MRC's mandate is to ensure "reasonable and equitable utilization" of the waters of the river.[26] Today, the member countries (Cambodia, Laos, Thailand, and Vietnam) have shifted the initial focus from gathering information and surveying the properties and potential uses of the river, to a view of the river mostly in terms of its potential for large-scale development projects. Some efforts are under way to move away from a focus on hydroelectric power potential to projects related to sustainable development and better management of the river's natural resources, but this shift conflicts with larger development needs. To foster sustainability, MRC's role and mission has attempted to promote good environmental governance principles, develop impact assessments, conduct risk analysis, and carry out conflict management and resolution strategies as needed.[27]

MRC runs programs related to all aspects of human interaction with the river. A list of current MRC programs is as follows: Agriculture and Irrigation Programme, Climate Change and Adaptation Initiative, Drought Management Programme, Environment Programme, Fisheries Programme, Flood Management and Mitigation Programme, Information and Knowledge Management Programme, Initiative on Sustainable Hydropower, Navigation Programme.[28] Over the past ten years or more, funding for MRC has come from the member states, and also from so-called development partners: the World Bank, the United States, Switzerland, Sweden, Australia, Belgium, Denmark, the EU, Finland, Germany, Japan, and Luxembourg.[29] This funding has made possible greater regional cooperation, monitoring, and communication of Mekong Basin conditions.

It is imperative to understand the significance of the Mekong to the region's population: in the period 2012–2015, fishing and aquaculture produced *an annual* 4.4 million ton catch, worth almost $17 billion (which is estimated to be 13 percent of the value of the world's total freshwater fish catch). The river also contributed $3 billion to Cambodia's economy in 2015 (18 percent of its GDP), and $1.51 billion to Laos (12 percent of GDP). In 2015 in larger countries, fishing from the Mekong brought in $5.74 billion (3.1 percent of GDP) to Vietnam, and $6.72 billion (about 1.8 percent of GDP) to Thailand.[30] Despite the economic and environmental importance of the shared use of the Mekong River, and despite great optimism that the post–Cold War period would usher in greater cooperation and mutual benefits, MRC is now facing serious obstacles and impediments to being able to fulfill its mission. Countries must find a way to protect and conserve the Mekong's resources, yet it is clear that continued use patterns coupled with other environmental stresses will lead to overuse and economic and human ruin. Countries and communities view the river in terms of their own self-interest.

Some of the successes and lessons of the early years of cooperation over the Mekong include the important finding that "the greater the international involvement in conflict resolution, the greater the political and financial incentives to cooperate. The pace of development and cooperation in the Mekong River watershed over the years has been commensurate with the level of involvement of the international community."[31] Early successful collaboration on mapping, studying, and evaluating the river's resources were supported by the UN and wealthy donor countries like the United States and France. This gave the lower riparians an incentive to cooperate. While there was little activity in the Mekong Committee in the 1970s to 1980s, hopes were high with the creation of the Mekong Commission that the earlier spirit of cooperation would be renewed and that there could be peaceful and equitable outcomes on how to use the river's resources. The newly created MRC exemplifies the proliferation of regimes and agreements in Asia in the 1990s to 2000s as countries sought to "make and keep agreements that incorporate jointly enacted rules, without debilitation, fear of free-riding, or cheating by others."[32] The signatory countries to MRC desire to "cooperate based on sovereignty equality and territorial integrity in the utilization and protection of the water resources of the Mekong River Basin."[33] MRC has no supranational authority; in other words, the organization cannot act on its own without approval of member states. MRC has a secretariat, located in Vientiane, which oversees the activities of the various programs listed above.[34]

The first Mekong River Commission summit was held in April 2010. Heads of state from the four member states met and renewed their commitment to the goals of the commission. They agreed in January 2011 to the Integrated Water Resources Management (IWRM) development strategy. This strategy incorporates goals of achieving higher levels of economic and social development. But MRC has not been a particularly successful organization and it is being presented with increasing tensions and points of contention; there is little reason for optimism that conditions will improve.

Current challenges are as follows: divergent priorities and plans for uses of the river (conflicts of national self-interest); a drop in funding and interest by previous donor countries; increased power and influence from China and the creation of a (possibly competing) new international organization, the Lancang-Mekong River Cooperation Mechanism; and, finally, a severe drought in 2016 that exacerbated all of the above stresses and dynamics. MRC, created to provide capacity building and mapping, was able to foster cooperation on these functional kinds of issues. Over time, as countries have had to contribute more resources, and be more willing to discuss how to share (and yes, limit) water and fishing rights, the organization has fallen short.

Divergent Priorities of Member States

The four member states view MRC differently and have different priorities for it. Thailand is the largest and most industrialized of the members and they may be less motivated to want to create and adhere to shared norms and rules over the Mekong. Thailand may be concerned that rules will prevent it from development projects that divert water from the river. Thailand is also less interested than other members in the consultative and institutional capacity that MRC can provide members. Thailand has a well-established and capable bureaucracy and is less interested in getting advice or support in developing these capacities. As an upstream state, Thailand is also less interested in the needs and fears of downstream neighbors in relationship to water flow targets and quotas (perhaps justifiably, Thailand claims that without Chinese participation water allocation discussions are meaningless).[35] As discussed below, Thai companies (banks, construction firms, and electric power companies) stand to benefit from dam projects in neighboring Laos.

Laos is the least developed country among the four, and 97 percent of its territory is within the Mekong River Basin. Despite the salience of the river to Laos's well-being, Ellen Bruzelius Backer finds that there are indications that "the Laotian government prefers to have the liberty to develop the Mekong tributaries according to its own preferences without having to adhere to regime recommendations."[36] Laos is of two minds on the utility of MRC. Given its lack of government capacity, it is willing to accept the resources offered by MRC, particularly on things like expert surveys, policy recommendations, data gathering, and implementation. However, Laos is reluctant to carry out the resulting policies that may hurt its economic interests. For example, it has refused to conduct comprehensive impact studies on the effects of logging on the water flow of the river.[37]

Laos is also at the heart of a significant conflict over a controversial dam project at Xayaburi, in northern Laos. Plans for the dam have been undergoing a consultation process (under the aegis of MRC) since 2010. Cambodia and Vietnam fear that the dam would adversely affect fisheries and reduce sediment flow needed for agriculture in their countries. In December 2011, the members of MRC agreed on the need for further study of the dam's effects on the environment and "the understanding was that no dam would be built until the study was completed."[38] The dam would generate electricity for the utility, Electricity Generating Authority of Thailand (EGAT), which would sell the power to Thai customers. In addition, financing for the dam is through several large Thai banks, and the project is being built by the Thai construction firm Ch. Karnchang.[39] The human, environmental, and economic costs will accrue elsewhere. There are estimates that the dam will block important fish migration patterns for 23–100 species of fish as well as destroy the river's complex ecosystem, affecting nutrient levels and agriculture as far downstream as the Mekong Delta in Vietnam. More than 200,000 people will be affected directly by being forced to relocate, or indirectly by having their livelihood destroyed (agriculture and fishing in areas near the dam will be radically altered).[40] Building the dam is also problematic because of the precedent being set. There are eleven more dams planned (nine in Laos) along the lower Mekong.

> Neighboring governments are concerned. According to a 1995 treaty, the four governments of Cambodia, Laos, Thailand, and Vietnam arc required to seek agreement before building any of the Mekong dams. The process is called "prior consultation" and facilitated by the Mekong River Commission. However, such agreement was never sought for the Xayaburi Dam. Despite requests from Cambodia and Vietnam, Laos and Thailand even refused to study the dam's transboundary impacts before beginning construction. The project's current environmental impact assessment only examines impacts 10 kilometers downstream from the dam site. Laos' engineer for the project, the Pöyry company of Finland, also claims that it can rely on unproven technologies to mitigate any harm the dam might cause. The full extent of the dam's impacts remains unknown.

> Construction activities began in late 2010, but Laos called these activities "preliminary work" for several years and even claimed at times that construction was delayed. On 7 November, 2012, Laos and Thailand held an official groundbreaking ceremony after claiming that Cambodia and Vietnam now support the project. At the MRC's Council meeting in January 2013, Cambodia, Vietnam, and the MRC's donor governments all continued to raise concerns about the project. Construction on the project is proceeding rapidly and much of the river has already been blocked off.[41]

Laos's deputy minister of energy, Viraphonh Viravong, made it clear (contradicting an earlier statement by the foreign minister) that work on the dam is proceeding without waiting for further studies of the economic or environmental impact.[42] Thai firms stand to profit from the project so we are unlikely to see protests from Thailand; that leaves Cambodia and Vietnam as the countries who are opposed to the project, but there is little indication that this will alter the plans for this dam and possibly others. Conflict over the dam demonstrates the weakness of MRC and the low level of willingness of member states to prioritize cooperation over local interests.

One would think that Vietnam would have a significant interest in the operationalization and functioning of MRC. Twenty percent of Vietnamese territory lies within the Mekong Basin and the delta in the south is agriculturally significant for the country. The delta region is home to 17 million people; they produce 90 percent of Vietnam's rice (much of it for export) and 53 percent of the shrimp and fish exports, and the area comprises 27 percent of the country's total GDP.[43] During the dry seasons, saltwater intrusion can ruin crops and farmland, as happened in 2016. Vietnam, in theory, is committed to cooperating within the framework of MRC, however, it has not always acted as such. Being upstream of Cambodia has not always made Vietnam mindful of its decisions about uses of the Mekong (as will become clear below). The Vietnamese government is sensitive and not particularly transparent about sharing information or being willing to compromise sovereignty within MRC.[44]

Eighty-six percent of Cambodia's territory surrounds the Mekong River Basin and millions of people rely on the river for fishing, farming, and transportation.[45] These people will be severely impacted by dam construction upstream. This makes Cambodia interested in getting MRC to be more robust. Cambodia is concerned about what can happen when countries upstream are not mindful of their actions. In 2000, Vietnam, without previously notifying Cambodia, opened the gates at Yali Falls Dam (70 kilometers upstream from the Cambodian border). The increased water flow resulted in unexpected floods in Cambodia that killed several people and did a great deal of damage.[46] Despite Cambodia's interest in MRC and how it is impacted by neighbors' actions on the

Mekong, it has not devoted significant resources or political clout to getting other member states to be more cooperative. Vietnam's decision to open the gates at Yali Falls could be construed as a violation of Cambodia's sovereignty and territorial integrity. While no Vietnamese troops were sent over Cambodia's borders, Vietnam's actions had lethal consequences in Cambodian territory, illustrating the close connection between human and national security.

Loss of Funding and Fragmentation

MRC is facing a significant budget drop. Since its inception, member states have borne only about 10 percent of the annual budget of the organization. Thus, the majority of the funding has come from the Asian Development Bank and the World Bank, and from major donors such as the United States, Switzerland, Sweden, Australia, Belgium, Denmark, the EU, Finland, Germany, Japan, and Luxembourg. In 2016, much of this financial support was withdrawn. Some countries have cut their support due to currency weaknesses (Australia, the EU countries, Denmark, and Japan), and others because of continuing softness of the global economy. Donor countries are also angry at Laos's continuation of the Xayaburi Dam project over objections of other member states. Finally, donors have concerns that money previously given has been mismanaged and subject to fraud. Donor states also would like to see MRC become financially self-sufficient, but there have been few efforts to move toward this. Donor countries supported MRC because it was a way to consolidate peace through functional cooperation in the region. If countries neither demonstrate the willingness to cooperate on uses of the Mekong, nor are they responsible with the funds, then major donors will be less willing to go forward in supporting the organization. For 2016, contributions from members were to go up by an additional 10 percent, but this is still not enough to cover the cost of the organization's work. MRC needs $65 million to fund the next five-year plan: "With $15 million coming from member countries, $9 million from an ongoing fund, the balance of $41 million will require external support from the donor community. . . . As of now, MRC has just $24 million it can count on."[47] Without funding, and given the other challenges that MRC faces, it is hard to see how well it can possibly function in the near term. Since MRC has common interests with the Asian Development Bank's GMS and since both have permanent bureaucracies, one would think that there would be strong coordination between the two organizations; however, that is not the case. At the state level, only Laos has the same point person (a minister in the prime minister's office) appointed to both organizations. In other countries different people, at different bureaucratic agencies, deal with the two organizations, and Sokhem Pech and Himesha Ranamukhaarachchi find that there is poor exchange of information between the organizations.[48] Given the weaknesses of the MRC, one might think that other organizations would be better positioned to address common concerns. ASEAN is the region's most comprehensive and developed organization, and it does address many of the issues related to shared use of natural resources.

Poor or almost nonexistent coordination between GMS and MRC personnel reflects the complex governance landscape surrounding the Mekong. These overlapping regimes are tasked with related issues and challenges around shared uses of the Mekong. The weak relationship found among government personnel and these regional organizations demonstrates conflictive fragmentation. The institutions and actors are hardly connected to each other; they have different decisionmaking procedures and conflicting norms, rules, and priorities along with different memberships.[49] In an ideal world these organizations would provide an example of synergistic fragmentation, where different committees of the regional organizations worked on different goals and programs to enhance cooperation and share outcomes of uses of the Mekong. But this is not what is happening on the ground.

ASEAN's Role

ASEAN would seem like a logical institution to address the kinds of regional problems and opportunities that cooperation over the Mekong reflects. ASEAN has an existing bureaucratic framework and membership that encompasses all Mekong states. ASEAN has become part of the network of organizations working to address uses of the Mekong. In the mid-1990s, ASEAN formed the ASEAN Mekong Basin Development Cooperation (AMBDC) forum to foster dialogue and cooperation among ASEAN countries and China over use of the Mekong. AMBDC also invites ADB and MRC to participate in its meetings; thus, this is the most comprehensive forum for relevant matters. Likewise, ASEAN has other ongoing programs that address water management and environmental issues, so there is institutional knowledge and a framework to address issues relating to the Mekong.

However, ASEAN has heretofore been more focused on political and security issues than concerns over natural resources and the environment. Also, in general, ASEAN is hampered by its consensus-driven mechanism for decisionmaking and its strict norms of noninterference.[50]

However, China could play a more forceful roll in organizations of its liking. To this effect, at the ASEAN + China summit meeting in November 2014, the idea was pitched to create the Lancang-Mekong River Cooperation Mechanism.

Truncated Membership and Competition from LMRC

China and Myanmar are not full members of MRC, but are observers. China is now clearly recognized as the Asian hegemon and, when it comes to almost any question about natural resource use and management, no discussion is complete without evaluating China's roll. This is even more true in the case of the Mekong River. The river originates in the Tanggula Mountains in Qinghai province, China. It flows through the Tibetan Autonomous Region, Yunnan province, then into Myanmar, Laos, Thailand, Cambodia, and Vietnam. Over the past thirty years as China's economy has boomed, so have its energy needs and level of industrialization. The area of the Mekong (called Lancang in China) within China is tailor-made for hydropower development; there are steep valleys and gorges.[51] Indeed, since 1994 China has built (or is currently building) six dams on the Mekong and there are plans for seven more.[52] (Eleven other dams are under construction or in the planning stages in other Mekong states, many with Chinese assistance and investment.[53]) China has signed on to terms of agreement with MRC for technical cooperation on sharing water-level information and control of the water flow. This agreement has been in place since 2004. China is, however, very much in the driver's seat on this. It has stated that its use of the Mekong within Chinese territory (with dam development) will be helpful to countries downstream, and it claims that there will be little or no environmental harm. These pronouncements have been met with skepticism from downstream neighbors and international experts.

In November 2014 at the ASEAN-China summit meeting, Chinese premier Li Keqiang proposed to establish the LMRC. The organization is designed to promote cooperation and consensus regarding political security, economics, and sustainable development as well as social and cultural exchanges. The goal is to promote connectivity, capacity cooperation, cross-border cooperation, water resource cooperation, and agriculture and poverty reduction cooperation.[54] These goals are similar to those of the GMS, which even has the same state members but with the additional element of including cooperation on water resources (not included in the GMS goals). It is worth noting here that member countries (with the key exception of Vietnam) are generally the states friendliest to

China. Cambodia, Laos, and Thailand have been China's most reliable friends in Southeast Asia, this would seem to auger well for LMRC cooperation and success.

Yet the recent history of China's behavior on water issues in the area does not look so good. In 2008, a flood in northern Laos and parts of Thailand were linked to China's Jinghong Dam, upstream in Yunnan. The dam started to generate power earlier that summer and China did not keep its neighbors informed about possible increased water flow from the dam's activities. In 2010, the opposite problem occurred. China's Xiaowan Dam created a large reservoir and, soon after becoming operable, record low water levels in the Mekong affected fish yields and water supplies for irrigation, drinking, and other uses.[55] In response to criticism, China began sharing dry season hydrological data (this was not an ongoing policy decision, but rather a one-off sharing of information) from two other dams, Manwan and Jinghong, and invited state representatives from neighboring countries to tour the Jinghong Dam.[56] China's dam (more aptly called mega-dam) projects on the Mekong dwarf all those downstream by a significant factor. The Xiaowan Dam is taller than the Eiffel Tower and produces 4,200 megawatts of electricity, and the Nuozhadu Dam has a 190-square-kilometer reservoir; either of these alone is larger than the combined current hydropower generating capacity of all the lower Mekong states.[57]

Since early 2015 multiple meetings were convened, and the organization was formalized in November 2015. China sees LMRC as part of its efforts to improve infrastructure and economic links between Southeast Asia and China. It fits well within China's larger development plans for power generation, access to natural resources, and its desire to build alliances or spheres of influence along its borders. A few months after the formation of the LMRC, China found itself in a position to help its fellow members. A longer-than-normal dry season, driven partly by climate change and partly by the El Niño cycle, resulted in significant drought conditions in the Mekong region. China offered to release water from upstream to help alleviate some of the problems. This "benevolence" demonstrates exactly what China hopes for out of the LMRC: that Chinese leadership will encourage countries to cooperate under a set of rules, norms, and procedures created by, and presumably set up to benefit, China. In return China may, if it feels so inclined, assist regional partners in times of need.

The weaknesses and fragmentation of the GMS and MRC have opened the door for China to fill a vacuum. With little funding and political will to allow the GMS and MRC to succeed, China has realized that it can design an alternative organization to promote its specific interests for the Mekong.

The 2016 Drought

In 2016, the Mekong River was at its lowest levels in 100 years (when record keeping began). We are clearly seeing the consequences of years of water overuse: a severe drought brought on by a harsh El Niño weather pattern (a weather phenomenon expected to bring below average rainfall, hotter temperatures, and overall drier conditions, all of which exacerbate the drought). The drought is causing economic hardship and stress up and down the Mekong River. It has also led to new migration patterns and tensions and significant threats to rice production, and thus threats to food security, in Vietnam, Thailand, and India. "Among the hardest hit areas are Vietnam's Mekong Delta (a rice bowl of Asia) and central highlands: 27 of Thailand's 76 provinces; parts of Cambodia; Myanmar's largest cities, Yangon and Mandalay; and areas of India that are home to more than a quarter of the country's massive population."[58]

> In Vietnam alone, nearly 160,000 hectares of farmland had been ruined by saltwater intrusion as of March 9 (2016), totaling $10.5 million in losses in the agriculture sector. Brackish water will likely damage an additional 500,000 hectares of paddy rice in the first half of 2016, a small but significant portion of the year's harvest. Thailand's rice production during the dry season has likewise been adversely affected in recent months, as has sugar crop production. Poor rice quality and lower yields in the Mekong and nearby Chao Phraya basins, coupled with an unplanted second offseason crop, have hindered projections. Harvests of poor quality sugarcane likewise resulted in lower yields. Since the first three months of the season, sugar production is already down 10 percent from the previous season, and even optimistic estimates predict the smallest crop in at least four years.[59]

Saltwater intrusion is a regular problem; however, because of the two-year drought from 2015–2016, conditions are especially bad, ruining groundwater and rice paddies as far as 90 kilometers inland. Under normal rainfall, Leocadio Sabastian, regional program leader for the International Rice Research Institute's office in Hanoi, explains that "the river's fresh water drives more saline water back to the sea."[60] This drought has contributed to declining reservoir levels and to the horrific forest fires in Indonesia this past fall and winter. It is also leading to food insecurity and a decrease in people's access to fresh drinking water, which can lead to an increase in water-borne diseases. Researchers with the International Center for Tropical Agriculture (CIAT) have shown the press and governments in the region maps that illustrate how water scarcity in the 2015–2016 drought, and longer-term trends associated with climate change, will put key crops at risk. Crops such as rice, cassava, maize, coffee, and cashew nuts in Vietnam, Cambodia, Thailand, and Laos are increasingly insecure. For example, supplies of rice are becoming dangerously low, and the government of Vietnam has approved $23.3 million in emergency funds to compensate farmers, provide water tanks, and so forth.[61]

> Not only does this affect the livelihoods of the Vietnamese farmers, but it also affects the biggest buyers of Vietnamese agricultural goods.
>
> "People in Indonesia and the Philippines will go hungry if the Thais and Vietnamese don't produce enough rice," Richard Cronin, director of the Southeast Asia Program at the Stimson Center in Washington, told Bloomberg. "This is a preview of the longer-term effect of development and climate change to the Mekong Delta."[62]

Despite China's water release in the spring of 2016, the People's Republic of China resists proposals for a water-sharing treaty among Mekong countries. There is unstated concern that China will continue to use its water control in a strategic way and as a potential political weapon. By controlling the flow of water for the other five countries in the Mekong region, China has significant leverage over its neighbors' behavior. China denies that it is stealing water or that its dams have contributed to shrinking water levels and recurrent droughts (or sporadic and unexpected flooding). Yet by continuing dam construction (both within China and through helping Thai firms in Laos build dams in Lao territory), it is all but assured to see such challenges continue and in all likelihood to worsen.[63]

Analysis

Discussion of natural resource development or management in the Mekong region reflects the complexity of transnational or transboundary issues. There are a multitude of interests and centers of power and decisionmaking that should be considered. As Pech and Ranamukhaarachchi detail, we must understand that there are six nations involved,[64] dozens of local authorities, hundreds of local communities and interest groups (including civil society groups and corporate interests) that have different economic, political, social, and economic objectives and unequal financial and technical capabilities. To understand cooperation or a lack of cooperation, we must have a way of understanding some of these dynamics and interests.[65] One way to understand this complexity is through the framework of networked governance. As we conclude below, riparian nations have not come up with a shared set of rules and treaties over water rights and usage of the Mekong. Civil society groups (both domestic and international) and community organizations have had little success convincing governments to take into consideration their concerns about environmental damage to the river, and the human costs of relocating whole villages due to dam construction, not to speak of the larger harm done to communities downstream from diverted water sources. Governments have sided with the interests of large construction firms and power companies seeking more hydroelectric power plants and dam construction. One part of the explanation is that states such as Laos, Cambodia, and Myanmar have poor track records in respecting human rights and human security as well as poor coordination between the national government and local governments. Much of this, however, is by design not just by default; national governments do not want to cede control to local level officials and there are few mechanisms for the people to hold leaders responsible for poverty and economic insecurity. Ruling parties and regimes are beholden to wealthy interests and are only weakly accountable to the people. Thus, it is easier to ignore, harass, and suppress activism that does not accord with state goals and priorities than to develop more cooperative and mutually beneficial relations between government and nongovernment actors, between national and local levels of the government, and between governments and regional bodies such as MRC and the GMS.

Looking specifically at how one or two countries approach regional governance of the Mekong, we can understand the difficulties embedded in trying to foster broader regional cooperation. In Cambodia, multitudes of government agencies and ministries are tasked with overseeing Cambodia's interest in the Mekong.[66] The primary government agency coordinating natural resources management is the Cambodia National Mekong Committee (CNMC), which is chaired by the minister of water resources and meteorology, with membership of seventeen other ministries (ranging from the Environment Ministry to Industry, Mines, and Energy, to Public Works, Tourism, Trade, etc.). The CNMC is charged with advising and assisting the government on all water-related matters as well as coordinating the different ministries with limited financial and human resources.[67] Staff at CNMC often complain of being bypassed or ignored by other ministries. Additionally, poor communication and interaction between officials at the national level in Phnom Penh and those at the district level contribute to low levels of coordination and effectiveness. Likewise, there seems to be little capacity for taking into account local concerns and community involvement in decisions that directly impact communities along the river.[68]

When one looks further at specific uses of the river like dam construction, and the related issues of resettlement, compensation, and environmental protection, the outlook is even more gloomy. In Cambodia, the Ministry of Industry, Mines, and Energy oversees policies and plans for dam construction and hydropower and it is the government agency charged with liaising with outside organizations such as the World Bank and ADB. The Ministry of the Environment is charged with implementing appropriate aspects of the Law on Environmental Protection and Natural Resource Management, which requires impact assessments of all development projects and activities. While the law dates from 1996, using environmental impact assessments is recent and there are few detailed guidelines for the process. However, there are several conflicts of interest embedded in the process:

- Responsibility for the impact studies lies with the project developer, making it suspect that it will have an interest in completing a thorough, independent, and honest assessment of the environmental and social impacts.
- The professional ability to complete a credible and scientifically sound impact study is weak.
- The time allocated for review is only thirty days, too short to properly evaluate conditions.
- There is little consultation with the public.
- There are few clear guidelines on completing social impact studies so that proper assessment can be done of the socioeconomic impact of new dams on affected communities.
- Monitoring and compliance are weak. This is true in Cambodia more generally, as there are few mechanisms in place for oversight and enforcement when standards and protocols are violated.
- Investment decisions often reflect political preferences (corrupt or cozy relationships between particular government officials and financing firms for the project) rather than economic or social benefits more generally.[69]

Controversy over dam construction and use of the Mekong exists at both the domestic level and the regional level. Within countries where agricultural producers have political clout, there is pressure for governments to work to preserve a more traditional flow and usage of the river. Where industrial interests are more powerful, such as those associated with power generation, construction, and banking, there may be a will to perform due diligence and environmental impact studies on the effects of dam construction on the wider region. The drought has also increased tension and potential conflict among neighboring states. There is the disagreement among Laos and its neighbors over dam construction, and another conflict between Thailand and Cambodia and Vietnam. In response to drought conditions, Thailand has diverted the Mekong River in its territory to tributary waterways to address severely parched conditions in the country. This has angered Vietnam, where officials have confronted Thai leaders over this. Cambodia has been more restrained. This may be because Cambodia feels that China will bail it out as needed. Thailand's actions did not include consulting with MRC on the diversion project; neither did it consult with the community and affected states.[70] The drought, dam activity, lack of funding, and lack of leadership among member states have combined to make a mockery of MRC. Even with recent restructuring of the organization, Vietnamese diplomat Pham Tuan Phan took over as leader from his Laotian counterpart, there seem to be few reasons for optimism that the organization will have any hope of playing a robust role in confronting the current challenges.

Laos provides another example of the multiplicity of layers of government and private actors involved in activities and uses of the Mekong. The government of Laos in particular is vigorously promoting hydroelectric power to propel economic development.[71] Helping Laos in this effort are private firms investing or seeking construction contracts from China, Thailand, Vietnam, Russia, and Malaysia. Laos's 1997 and 2010 Law on Electricity are the critical laws guiding the development of dam projects. These pieces of legislation lay out which government agencies are responsible for which elements of administration, inspection, and obligations concerning power plants. The laws do not cover construction or the infrastructure development needed to dam bodies of water and get the power plants up and running. The Ministry of Energy and Mines (MEM) has leadership over private and foreign investment for hydropower development, and its decisions are to be governed by the National Policy on Environmental and Social Sustainability of the Hydropower Sector of 2007. To comply with this statute, a full environmental impact assessment is supposed to be carried out. However, the policy does not lay out institutional arrangements for projects to ensure sustainability or which agency should coordinate the different government bodies involved. MEM is the lead bureaucracy for dam development, but two other agencies (the Ministry of Natural Resources and Environment and the Water Resource and Environment Administration) would in theory be the ones to coordinate environmental and socioeconomic impact from building the dams. The laws mentioned above are vague and not specific enough in how impacts should be measured and how coordination among agencies should be carried out.[72] Other potential problems are as follows:

- The environmental impact assessment process is complicated and stretches across multiple agencies. So, developers are incentivized to bypass the process.
- It is not always clear which guidelines apply to which projects.
- There is no certification program for impact assessment consultants, so quality is spotty.
- Staff in government agencies are minimally trained and relatively inexperienced.
- Staffing levels are low in many of the ministries.
- Once national-level agencies approve a project, provincial-level officials must oversee implementation, and these officials have limited financial, technical, and professional experience.
- There have been only minimal attempts to allow and foster participation of communities that will be affected, and weak accountability to these citizens.[73]

In order for international or regional organizations to work, several elements must be in place. Member states need to agree on goals, rules, procedures, and agendas. Clearly, states along the Mekong have not all agreed to these. Then, even when there is agreement, states must be able to do what they say they are going to do. The thumbnail sketches of Cambodia's and Laos's bureaucratic structures to deal with hydropower projects illustrate that government agencies have limited capacity to handle complex technical projects like dam construction. Implementation and legal provisions are weak, subject to centralized interference and politicization (and opportunities for corruption); there is a lack of transparency and citizen protection or input; and the ability to execute at the local level remains weak. What this means is that countries lack rules and safeguards to regulate dam construction so as to protect workers and local communities from harm. Despite provisions from MRC and environmental treaties signed by member states, countries such as Laos and Cambodia have not done sufficient work to assess the environmental, social, and economic impact of proposed hydroelectric projects. Even basic technological resources (e.g., computers, software, Internet access, and monitoring devices) are lacking at the local level.[74] Regional organizations such as the Mekong River Commission rely on development decisions by agencies within sovereign riparian countries. MRC aims to help member states understand and plan for the long-term impact of dam construction, but MRC and other regional organizations and legal provisions in environmental treaties can be effective only if states want to comply with them and if states have the necessary capacity to do so. Countries have interpreted the authority and expectations of MRC differently, and there are varying levels of support and compliance for the arrangement and agreements. For example, MRC requires that all hydropower dams proposed for the Mekong must undergo proper procedures for "Notification, Prior Consultation, and Agreement."[75] While the government of Laos declared that it had completed this procedure for the dam at Xayabouri, other nations have charged that it has not done nearly enough to examine and notify all effected parties of the impact. In sum, there are grave problems at all levels of government and society. One cannot meaningfully discuss networked governance or problems of fragmentation when so little coordination happens.

At the international level, there are global efforts to help countries cooperate with their neighbors over water rights. The UN Convention on the Non-Navigational Uses of International Watercourses, approved in 1997, requires (in Article 5) that

> nations utilize international watercourses in an equitable and reasonable manner with a view to attaining optimal and sustainable utilization and benefits consistent with adequate protection in the watercourse. Article 5 also provides that watercourse nations shall participate in the use, development, and protection of an international watercourse in an equitable and reasonable manner. The right to participate includes both the right to utilize the watercourse and the duty to cooperate in its protection and development.[76]

Additionally, the Helsinki convention is also an agreement on river cooperation; it started as an agreement among European states under the UN Economic Commission for Europe, but has now been expanded to allow any nation to join. The Helsinki convention goes beyond the abovementioned UN convention with principles about negotiating deals on water sharing.[77] None of the Mekong riparian states are signatories to either of these global agreements. Further, China has asserted that it has "indisputable territorial sovereignty over those parts of international watercourses that flow through its territory."[78] So, there are international treaties, norms, and mechanisms to facilitate cooperation and the provision of the river as a public good. But if states are unwilling to sign on to legal frameworks to help achieve this, then there is little reason for optimism that fair, reasonable, and equitable sharing of the Mekong's resources is imminent.

Conclusion

There are significant water challenges in Southeast Asia. Countries are facing short- and long-term challenges in how they manage and allocate increasingly scarce water supplies. Cooperation over water in the Mekong is one of these necessary measures. It behooves all riparian states to come to agreement on what constitutes fair use of the river's resources. Even with a multitude of regional organizations, and international treaties created specifically to do this, countries are unable to cooperate in meaningful and comprehensive ways. Unwillingness to join global frameworks for cooperation make those treaties useless in this situation. Regional institutions themselves are weak, underfunded, and suffer from a lack of leadership. Lack of Chinese membership in MRC, along with China's new alternative organization, weakens MRC's ability to function as intended. In theory, LMRC could be a promising tool for achieving a fair and shared use of the river. Created by China, there is a chance that China could play a constructive leadership role in fostering greater cooperation within the organization. However, China has made it clear, heretofore, that it has no intention of giving up even a hint of sovereignty to this or any other similar association. Instead, China views LMRC as an effective way to exert its interests on neighboring countries, certainly to its own benefit, while claiming that the organization will benefit all member states. In other words, China is not interested in providing public goods, as hegemonic stability theory posits that an effective hegemon should. China is interested in maximizing its own power, effectively denying downstream states the ability to be able to provide public goods for their citizens through the continued use of the Mekong. Cooperation over the Mekong is a clear collective action problem. But without China's leadership, it is unlikely to happen. Even if regional organizations were to play a more constructive and influential role, it is unclear if member states along the Mekong are in a position to comply with the technical and oversight mechanisms necessary to protect the river for long-term use and sustainability. In the case of lesser-developed countries, such as Cambodia and Laos, weak government capacity (particularly at the local level), high levels of corruption, and little willingness or tolerance for community and NGO involvement make it hard to see where greater cooperation could be forthcoming.

Notes

1. John Brandon, "Asia: The World's Most Water-Stressed Continent" (Washington, DC: Asia Foundation, March 21, 2012).

2. Alison Smith, "Gone Fishing," *Southeast Asia Globe,* January 24, 2014.

3. The Asian Development Bank's report, "Water Rights and Water Allocation: Issues and Challenges for Asia" (Manila, Philippines: Asian Development Bank,

2009), details domestic and cooperative efforts in Southeast Asia to codify (or not) water rights within and between countries. Some countries, such as Thailand, have few domestic laws that address licensing, use, and rights for surface water, groundwater, and so forth. The Philippines, Laos, and Vietnam have water codes addressing these issues of resource allocation. Overall in Southeast Asia, there is significant variation in how well countries' legal systems have even begun to address use, management, and conservation of this vital natural resource.

4. Brahma Chellaney, "In Asia, Governments Struggle to Deal with Water Crisis," *Today Online,* April 28, 2016, http://www.todayonline.com/world/asia/asia-govts-struggle-deal-worsening-water-crisis.

5. Asian Development Bank, "Asian Water Development Outlook: Measuring Water Security in Asia and the Pacific" (Manila, Philippines: Asian Development Bank, 2013).

6. Brandon, "Asia," 2.

7. Ibid.

8. Ibid.

9. Sierra Ewert, "Eating the Last Drop: Changing Diets in Asia Challenge Future Water Security" (Washington, DC: Asia Foundation, March 21, 2012).

10. Ibid.

11. Ibid.

12. Jacqueline Torti, "Floods in Southeast Asia: A Health Priority," *Journal of Global Health*, 2, no. 2 (December 2012), https://www.ncbi.nlm.nih.gov/pmc/articles/PMC3529313/

13. "How to Move Flood Water Through Bangkok," *IRIN News,* June 19, 2012.

14. Deden Rukmana, "Recent Floods in Jakarta Are Strong Evidence that Urbanization Must Be Reduced," *Indonesia's Urban Studies Blog,* February 21, 2007, http://indonesiaurbanstudies.blogspot.com.

15. Jacqueline Torti, "Floods in Southeast Asia: A Health Priority," *Journal of Global Health* 2, no. 2 (2012): 1–6.

16. Sokhem Pech and Himesha Ranamukhaarachchi, "Water Governance Systems in Cambodia, Lao PDR and Vietnam" (Vientiane, Laos: Mekong Program on Water Environment and Resilience, 2013).

17. Mukand S. Babel and Shahriar M. Wahid "Freshwater Under Threat: Southeast Asia" (Nairobi, Kenya: UN Environment Programme, 2009), 32–33.

18. Jeremy Bird, Wouter Lincklaen Arriens, and Dennis Von Custodio, "Water Rights and Water Allocation: Issues and Challenges for Asia" (Manila, Philippines: Asian Development Bank, 2009).

19. Pech and Ranamukhaarachchi, "Water Governance Systems in Cambodia, Lao PDR and Vietnam," 18–20.

20. Aaron T. Wolf and Joshua T. Newton, "Case Study of Transboundary Dispute Resolution: the Mekong Committee" (Eugene: Oregon State University Program in Water Conflict Management and Transformation, 2000).

21. Ibid.

22. Ibid.

23. Ibid.

24. Mekong River Commission (MRC), "Mekong River Commission History," 1995, www.mrcmekong.org.

25. Apichai Sunchindah, "The Lancang-Mekong River Basin: Reflections on Cooperation Mechanisms Pertaining to a Shared Watercourse," Working Paper Policy Brief No. PO13-01 (Singapore: S. Rajaratnam School of International Studies (RSIS) Centre for Non-Traditional Security (NTS) Studies, February 2013).

26. MRC, "Agreement on the Cooperation for the Sustainable Development," Article 5, April 5, 1995.

27. Sunchindah, "The Lancang-Mekong River Basin."

28. Mekong River Commission Portal, "MRC Annual Report 2014," March 31, 2015, http://www.mrcmekong.org/assets/Publications/governance/MRC-Annual-Report-2014.pdf.

29. Information on funding can be found at the Mekong River Commission website, "Development Partners and Partner Organizations," accessed September 2, 2017, www.mrcmekong.org/about-mrc/development-partners-and-partner-organisations/.

30. Luke Hunt, "Mekong River Commission Faces Radical Change," *The Diplomat,* January 22, 2016.

31. Wolf and Newton, "Case Study of Transboundary Dispute Resolution," 31.

32. Robert Keohane, Peter Haas, and Mark Levy, eds., *Institutions for the Earth: Sources of Effective International Environmental Protection* (Cambridge: MIT Press, 1993), 19.

33. MRC, "Agreement on the Cooperation for the Sustainable Development," Article 4.

34. MRC is not the only regime involved in overseeing the use of the Mekong. The Asian Development Bank started a program, the Greater Mekong Subregion Economic Cooperation Program (GMS). The purpose of this organization is to foster infrastructure development along the Mekong, but there are environmental elements to the plan and there is overlap with MRC. For more on both the Asian Development Bank's and MRC's activities, see Ellen Bruzelius Backer, "Research Paper Insights: The Mekong River Commission: Does It Work, and How Does the Mekong Basin's Geography Influence Its Effectiveness?" Lysaker, Norway: Fridtjof Nansen Institute, 2007.

35. Ibid., 39.

36. Ibid., 40.

37. Ibid.

38. "Lies, Dams and Statistics," *The Economist*, July 26, 2012.

39. International Rivers Organization, "Xayaburi Dam" (Berkeley, CA: International Rivers Organization, April 2014); "Xayaburi Dam Timeline of Events" (Berkeley, CA: International Rivers Organization, April 2014), www.internationalrivers.org/campaigns/xayaburi-dam.

40 International Rivers Organization, "Xayaburi Dam"; International Rivers Organization, "Xayaburi Dam Timeline of Events."

41. International Rivers Organization, "Xayaburi Dam"; International Rivers Organization, "Xayaburi Dam Timeline of Events."

42. "Lies, Dams and Statistics."

43. Backer, "Research Paper Insights," 41.

44. Ibid., 42.

45. Ibid., 41.

46. Ibid.

47. Hunt, "Mekong River Commission Faces Radical Change," 2.

48. Pech and Ranamukhaarachchi, "Water Governance Systems in Cambodia, Lao PDR and Vietnam."

49. Frank Biermann, Philipp Pattberg, Harro van Asselt, and Faroborz Zelli, "The Fragmentation of Global Governance Architectures: A Framework for Analysis," *Global Environmental Politics* 9, no. 4 (2009): 14–40.

50. Sunchindah, "The Lancang-Mekong River Basin," 6.

51. Milton E. Osborne, *River at Risk. The Mekong and the Water Politics of China and Southeast Asia* (Sydney, Australia: Longueview Media for Lowy Institute for International Policy, 2004).

52. Shannon Tiezzi, "Facing Mekong Drought, China to Release Water from Yunnan Dam." *The Diplomat*, March 16, 2016.

53. "China Drives Water Cooperation with Mekong Countries," *The Third Pole,* February 1, 2016.

54. Ministry of Foreign Affairs, People's Republic of China, "Five Features of Lancang-Mekong River Cooperation" (Beijing: Ministry of Foreign Affairs, People's Republic of China, March 17, 2016), www.fmprc.gov.cn/mfa_eng/zxxx_662805/t1349239.shtml.

55. "China Drives Water Cooperation with Mekong Countries."

56. Ibid.

57. Brahma Chellaney, "China's Water Hegemony in Asia," *Washington Times,* April 26, 2016.

58. Ibid.

59. "Defining the Geopolitics of a Thirsty World: The Politics of the Mekong River" (Austin: Stratfor, April 11, 2016). www.waterpolitics.com.

60. Christina Larson, "Mekong Megadrought Erodes Food Security," *Science,* April 6, 2016.

61. Ibid.

62. Elena Holodny, "Vietnam Is Being Crippled by Its Worst Drought in Nearly a Century," *Business Insider,* April 27, 2016.

63. Chellaney, "China's Water Hegemony in Asia."

64. In addition to the riparian states, there are also interests at stake for countries that have invested money in the region, including Western countries who previously funded the MRC as well as private funds from Korea, Russia, Malaysia, Kuwait, Qatar, and India. Pech and Ranamukhaarachchi, 2013.

65. Pech and Ranamukhaarachchi, "Water Governance Systems in Cambodia, Lao PDR and Vietnam."

66. Government of Cambodia, "Ministry of Water Resources and Meteorology—Department of Meteorology Cambodia: About Us," accessed September 2, 2017, http://www.cambodiameteo.com/articles?menu=115&lang=en.

67. For information on laws impacting environmental issues and resource development, see the "Open Development Initiative," a project of the East-West Management Institute, 2015, http://ewmi.org/content/Open-Development-Initiative; for information regarding Cambodia specifically, see "Cambodia Open Development," 2015, https://cambodia.opendevelopmentmekong.net/laws.

68. Pech and Ranamukhaarachchi, "Water Governance Systems in Cambodia, Lao PDR and Vietnam."

69. Ibid.

70. Luke Hunt, "Drought Fans Tensions Along the Mekong," *The Diplomat,* March 12, 2016.

71. The Lao People's Democratic Republic would like to be a regional energy supplier and to use revenue from selling energy for broader development goals. See the government's perspective: Government of Cambodia Ministry of Mines, "Hydropower in Lao PDR," 2014, www.poweringprogress.org/new/2-uncategorised/3-hydropower-in-lao-pdr.

72. Pech and Ranamukhaarachchi, "Water Governance Systems in Cambodia, Lao PDR and Vietnam."

73. Ibid.
74. Ibid.
75. Mekong River Commission for Sustainable Development, "Official Prior Consultation Process of the Pak Beng Hydropower Project Kicked Off," January 13, 2017, http://www.mrcmekong.org/news-and-events/news/official-prior-consultation-process-of-the-pak-beng-hydropower-project-kicked-off/
76. "Law, International Water," *Water Encyclopedia*, accessed October 19, 2017, http://www.waterencyclopedia.com/La-Mi/Law-International-Water.html.
77. Fred Pearce, "A Global Treaty on Rivers: Key to True Water Security," *Yale Environment 360,* November 19, 2012, http://e360.yale.edu.
78. Ibid., 2.

5

Health Security: Avian Influenza and Disease Protocols

Diseases occur without concern for borders or national sovereignty, and pathogens cross state boundaries easily and with great regularity. Clearly, the spread of new and deadly diseases is a transboundary problem. However, the spread of infectious diseases is not consistently viewed as a national security threat. It is only when there is an outbreak of unusual size and severity, or an outbreak of a new and more deadly disease not easily treated or prevented, that states view health problems as a security threat. States are the primary providers of public health measures yet, for many health issues, states need to work together to cope with and prevent outbreaks of serious diseases.[1] One only need to think of the Ebola outbreak in West Africa, or fears about the Zika virus in 2016, to be quickly reminded how a deadly disease can become a regional and global problem. Regional and international organizations and agencies are deeply involved in working with states to study, prevent, treat, and control infectious disease outbreaks. Organizations and agencies at the forefront of dealing with infectious diseases include the World Health Organization, the Food and Agriculture Organization, the World Organisation for Animal Health (OIE), and the US Centers for Disease Control.

The magnitude of threat to human and economic well-being from diseases is high. A disease threat anywhere can mean a threat everywhere. *Global health insecurity* is defined as vulnerability from:

- the emergence and spread of new microbes;
- the globalization of travel and trade;
- the rise of drug resistance; and

- the potential use of laboratories to make and release—intentionally or not—dangerous microbes.[2]

Thus, *global health security* provides protection from all of these infectious disease threats. International health regimes, such as the WHO, are the world community's most robust mechanisms for cooperation. States have agreed to share information and follow proscribed protocols, but these international organizations can be effective only to the extent that local partners—including national policymakers, public health officials and bureaucrats, local populations, and local officials—are willing and able to use effective strategies to address the problem. This is a perfect example of the ideal of cooperation at work in complex and hybrid governance. Yet Southeast Asian states demonstrate how challenging this cooperation can be and how difficult it is to achieve effective results in combating disease outbreaks.

One of the most essential elements of human security is public health. Citizens need access to information on health and disease, and they need systemic networks of health care providers and treatment. The emergence of new and deadly diseases often triggers misunderstanding, panic, and anger in citizens. Governments are forced to address disease epidemics or face the potential ire of their citizens. Information, prevention, and treatment of newly emerging diseases is essential, not just for citizens and governments in one country, but globally since disease outbreaks are easily spread across borders. Thus, human security is linked to state security and international security as well; states fear disease outbreaks as a risk to human well-being and to regime stability. As with other collective action problems, countries want the benefits of the assistance from regional and global health systems, but they do not always want to contribute to the costs. Public health systems that can monitor and treat populations for disease outbreaks are a key component to human security and are an example of states' abilities to provide public goods. But in looking at the problem of avian influenza in Southeast Asia, we find that many countries come up short. The example of how countries in Southeast Asia have dealt with avian influenza is interesting because states do indeed see it as in their interest to protect their citizens from disease outbreaks; nonetheless, governance problems, bureaucratic jockeying, and political competition have hindered abilities to effectively address health threats.

We chose to look at avian influenza in this chapter because it has posed significant challenges to the region in providing a robust public health response and because the conditions that lead to an outbreak of the disease are omnipresent. Thus, a new outbreak can occur at any time. Countries in Southeast Asia have struggled to create and bolster public health networks to increase access to health care and address threats to public health. Across the region, the risk of a new major outbreak of avian influenza looms large. H5N1, a strain of avian influenza, also referred to in this chapter as highly pathogenic avian influenza (HPAI), is transmitted by both wild birds and livestock. Humans are vulnerable to the disease when they handle or consume sick or infected birds. All of the factors that could produce a regional or even a global pandemic are present.

Over the past several years, there has been a push to rethink how countries and the global community deal with emerging infectious diseases. Efforts have been made to securitize disease outbreaks and responses. With the outbreak of SARS in 2002,[3] and with Ebola in 2014–2015 and Zika in 2016, there has been a rise in the use of security language linked to public health crises and an increase in media coverage and funding to address the spread of deadly diseases. We see health officials and reporters using terms like "imminent danger" and "deadly threats."[4] Officials in government and in the private sector now see the spread of disease as a public health concern, an economic concern, and as a security concern. Along with these changes has come an expansion of what global health organizations such as the WHO are called on to do and increased power for them to carry out these functions. This increase in global health governance could be useful in preventing the spread of diseases, but it is not without controversy. While no country wants to deal with an outbreak of a deadly pathogen, there can be disagreement and controversy between individual countries' priorities and preferences for their interaction with the global community, and the priorities of the international organizations themselves, as well as the preferences and goals of wealthy countries that fund and have significant influence within these institutions. There is also tension between state- and local-level measures used to address disease outbreaks when and where they occur. So, there are multiple axes where conflicting interests impact surveillance and efforts to combat infectious diseases.[5] When looking specifically at H5N1, we see a clear example of this: there are regional and global interests in preventing the spread of the EID, but there are differences in priorities in addressing outbreaks of the disease among countries within the region, and differences between donor and recipient countries in how to most effectively use the aid to combat disease. Experiences in Thailand, Vietnam, Cambodia, and Indonesia are illustrative of what can be done well and of what obstacles still exist in how we confront and deal with common problems like avian influenza. Each of the four case studies demonstrates something different about how countries cope with deadly disease outbreaks and how developing countries respond to increased efforts at global health governance.

This chapter unfolds as follows: first, we look at how Thailand, Vietnam, Cambodia, and Indonesia have responded to H5N1 outbreaks since 2004. Next, we examine efforts of a range of different actors involved in addressing the disease: international organizations and large donor nations have committed financial and scientific resources to assist the region in these efforts. Then, we explain how some of these efforts have been at odds with what happens within the countries themselves. The actors include policymakers in state capitals, bureaucracies charged with implementing policies and protocols, local officials, NGOs and private interest groups, groups of citizens (e.g., families with backyard poultry), and larger commercial producers. In the best-case scenarios, there is a scalar or ladder-like relationship going from the local to the global and vice versa. The WHO mandates certain reporting standards and offers guidelines for best practices for dealing with disease outbreaks; national governments have to decide how to comply with these requirements, and then government and nongovernment actors must carry out the work on the ground at the local level. We analyze what the prospects are for greater cooperation within states among different actors or stakeholders, between states and international organizations, and among states within Southeast Asia.

There are three fundamental problems that can be discerned by looking at Southeast Asia's responses to HPAI: first, there are different priorities at stake from international organizations and from states themselves. International organizations, such as the FAO, the WHO, and the global community at large, want to see deadly diseases confined and eradicated so that fewer countries and populations are impacted and so that larger economic networks are not disturbed. Individual countries may have different interests. As we show in the case of Cambodia and Indonesia, countries may prioritize getting access to vaccines over implementing greater culling measures. The goal of culling is to stop the spread of the virus rather than just treating it when it occurs. Other states may shy away from vaccinating poultry because of export concerns and, despite seemingly reasonable intentions, countries may opt not to follow best practices for preventive measures.

Related to these differences in priorities, we also find that interest group politics affects how well countries will address and combat HPAI. Large agricultural producers in Thailand and Indonesia may work to protect their own interests by making policy decisions more about economic and political power than public health concerns. Likewise, we find in the case of Cambodia and Indonesia that the ability of the central government to carry out policy is vastly more complicated under decentralized political control in Indonesia and also is problematic in Cambodia where the physical infrastructure as well as health and education infrastructure is quite poor.

Overview of H5N1

Avian influenza, or highly pathogenic avian influenza, commonly referred to as bird flu, is a viral infection found in a variety of bird species. Most bird flu viruses do not infect humans; however, two strains of avian influenza (H5N1 and H7N9) do cause serious infection in humans. The majority of cases of avian influenza have resulted from contact with infected live or dead poultry. There is no evidence that the disease can be transmitted to people through properly cooked food. However, the CDC is now stating that it is possible that very close contact with an infected person may transmit the disease from person to person: "Currently, HPAI Asian-origin H5N1 virus does not transmit efficiently from person to person. Some cases of limited, non-sustained human-to-human transmission have likely occurred."[6] Such transmission has occurred only through prolonged and close contact within families. If human-to-human transmission is possible, then the possibility of a pandemic is much stronger.

H5N1 first infected humans in 1997 in an outbreak in Hong Kong.[7] Significant outbreaks in humans have occurred each year in countries throughout East Asia and Southeast Asia from 2003 to the present.[8] Table 5.1 provides information on rates of avian influenza in Southeast Asia. Indonesia has by far the largest number of cases, and it has done a relatively poor job in combating the disease. Cambodia, Thailand, and Vietnam have also had to combat outbreaks, and we discuss differences in how each country has done so.

While the virus has now been found in poultry throughout Asia, Europe, and Africa, the risk factors for transmission to humans are especially prevalent in Southeast Asia. The greatest risks for exposure to avian influenza can include direct or indirect exposure to infected live or dead birds, or contaminated environments such as poultry farms, husbanding of wild birds, and live bird markets. There are three properties of avian influenza that make it worrisome as a vector for a pandemic: the disease infects humans, people do not have a natural resistance to the virus, and it is highly lethal. In Southeast Asia small-scale food production is common and consumption of poultry is widespread. Most poultry, including chickens as well as domesticated and wild ducks and geese, are raised by small poor farmers with little knowledge of biosecurity measures. Birds and humans tend to live in close proximity and open air markets are common. Poverty, malnutrition, and poor health exacerbate illness, so morbidity rates of the disease are higher in this region than they might be elsewhere. As of yet, the disease cannot be easily passed from human to human, but only one genetic mutation is needed for this to become possible.[9]

Table 5.1 Confirmed Human Cases of Avian Influenza (H5N1)

Year	Cambodia	China	Indonesia	Laos	Thailand	Vietnam
2003	0	1	0	0	0	3
2004	0	0	0	0	17	29
2005	4	8	20	0	5	61
2006	2	13	55	0	3	0
2007	1	5	42	2	0	8
2008	1	4	24	0	0	6
2009	1	7	21	0	0	5
2010	1	2	9	0	0	7
2011	8	1	12	0	0	0
2012	3	2	9	0	0	4
2013	26	2	3	0	0	2
2014	9	2	2	0	0	2
2015	0	6	2	0	0	0
2016	0	0	0	0	0	0

Source: Data from WHO Regional Office for Southeast Asia, "Surveillance and Outbreak Alert: Avian Influenza in 2013," updated 2017, www.searo.who.int/entity/emerging_diseases/topics/avian_influenza/en/; WHO Global Influenza Program, "Influenza," data as of January 6, 2015, http://www.who.int/influenza/en/; WHO, "Cumulative Number of Confirmed Cases for Avian Influenza 2003–2017," July 25, 2017, http://www.who.int/influenza/human_animal_interface/2017_07_25_tableH5N1.pdf?ua=1.

Note: Data for Malaysia is not available from WHO.

International health officials and agencies agree that there are several measures necessary for curtailing the spread of H5N1 and preventing outbreaks of the disease in humans. There needs to be close monitoring and oversight of poultry; areas of infection must be quarantined; and sick birds or birds suspected to be sick should be vaccinated or culled. Farm-raised poultry should have some measure of isolation from wild birds, and there should be limitations on the movement of poultry. There should be comprehensive testing and treating for exposed humans as well as public information campaigns. And public health institutions need to stockpile antivirals for human use.

There is a significant gulf in Asia in how well countries are prepared to handle disease outbreaks, and this has been demonstrated by their efforts to deal with avian influenza. Japan and Taiwan have extensive stockpiles of antivirals and oversight of poultry. Singapore has antiviral stockpiles able to dose 15 percent of the population, and it has enhanced surveillance and preparedness plans. Outbreaks in Japan and Taiwan were quickly controlled, and they are now free of the disease. Thailand and Vietnam were hard-hit and had problematic initial responses to the disease. But both countries changed how they tackled avian influenza and later received praise from health officials for better containing the disease and for working with international agencies. Likewise, donor governments have devoted greater resources and stepped up their responses to these countries over time.[10] Cambodia and Laos have poor infrastructure, both physical infrastructure (e.g., roads and health clinics) and communications infrastructure (for education and public awareness campaigns), so there are significant gaps in responses and outbreak capabilities. Cambodia and Laos are also extremely poor countries, so villagers have eaten sick birds because of a lack of greater food options.[11] Indonesia demonstrated a poor initial response, and much has been written about Minister of Health Siti Fadilah Supari's claim of viral sovereignty and decision to close Naval Medical Research Unit No. 2 (NAMRU-2) in 2008. NAMRU-2 was established in 1970 by the US Navy, and its work was carried out in cooperation with and under the auspices of the Indonesian Health Research and Development Agency (under the purview of the Ministry of Health). NAMRU-2 was one of the centers responsible for monitoring and studying avian influenza throughout Southeast Asia.[12] Despite Supari's politicized and publicized decision, discussed further below, Indonesia's response to avian influenza has improved over time. It is still weak and less robust than that of its neighbors in Thailand and Vietnam.

Why, despite similar risk factors and large outbreaks as well as time, focus, and funding from developed nations and international organizations, have we seen differences in the responses in Thailand, Vietnam, Cambodia, and Indonesia? The reasons for these differences can be explained by vastly different political and technical conditions and capabilities in the four countries. The most significant differences fall into two categories. The first category involves complex government relations, namely the dynamics between central governments and three different sets of actors: with other government bureaucracies, with local officials, and with the international community. The international community is comprised of international organizations, wealthy donor countries, and aid organizations. The second category is the role of powerful interest groups within each country and how these interests affect policy on HPAI and its implementation. These differences, and the hybrid nature of these actors, produce noticeable variation in how well the four countries discussed here have dealt with HPAI, and these differences tell us a lot about a country's ability or inability to provide public goods like disease monitoring. Countries certainly see it as in their interest to prevent and treat disease outbreaks, but political and technical problems hinder this.

Risk Factors for HPAI

Southeast Asia, unfortunately, demonstrates a number of conditions that make it at high risk for bird flu outbreaks:

- There is large demand for locally produced poultry and poultry products;
- A high proportion of poultry is raised and sold in poor conditions that provide little protection against HPAI;
- Significant numbers of birds in infected areas do not display symptoms of the virus and, thus, farmers are reluctant to take precautionary measures such as culling or vaccinating;
- The veterinary and public health systems throughout the region are weak, which makes reporting on the disease and combating it problematic;
- There are low levels of commitment toward addressing the problem within the poultry sector, governments, and the public.

Despite these risks and obstacles, we do see examples in the region of countries that have done a fairly good job in trying to address disease outbreaks.

State Responses to H5N1

Thailand

Thailand's response to avian influenza demonstrates how politics impacts disease response. And it shows how a developing country is able to tackle this kind of threat. After being hit in 2003–2004 with a significant outbreak of avian influenza that resulted in seventeen human cases of H5N1, Thailand was able to successfully implement a series of measures that saw immediate results: the number of human cases shrank to five in 2005, then three in 2006, and there have been zero additional human cases reported since then. By strengthening infrastructure and public awareness, as well as building confidence and cooperation from different sectors of society, Thailand has successfully avoided larger numbers of fatalities from H5N1.[13]

The 2003–2004 outbreak hit Thailand on the heels of economic and political upheaval. Business tycoon Thaksin Shinawatra had won the January 2001 election in a landslide. He was the beneficiary of a new, although short-lived, constitution approved in the middle of the 1997–1998 financial crisis. Thaksin's widespread electoral support illustrated high levels of his populist appeal from rural and semirural Thais as well as from the Thai business class. His leadership style was centralized and autocratic.[14] Also in the late 1990s to early 2000s, Thailand's poultry market and exports were growing rapidly. By 2003, Thailand was the world's fourth-largest poultry exporter and exports were estimated to contribute $1.21 million (48.4 billion baht) to the Thai economy (exports represented about 40 percent of poultry production).[15] Local Thai consumption of poultry is also quite high. Poultry production takes several forms: backyard chicken and duck production (the largest sector), small- to medium-sized commercial farms, large-scale farms, and vertically integrated farms that deal with all stages of the poultry industry from hatcheries to postslaughter processing (these are small in number, but huge producers).[16]

The appearance of avian influenza in 2003 was not surprising, but there was little response from the government. By early 2004, domestic and international pressure increased, so the government launched a systemic national surveillance program and began extensive measures to limit and prevent further outbreaks.[17] Prime Minister Thaksin convened high-level government meetings with key ministries, including the Ministry of Agriculture and Cooperatives, Ministry of Public Health, Ministry of Commerce, and Ministry of Foreign Affairs (the latter two reflecting the importance of Thailand's exports). The measures enacted followed the recommendations of the FAO, the WHO, and the OIE: comprehensive culling of infected birds, cleaning and disinfecting of the equipment and facilities in use with infected birds, restrictions on movement of poultry into and out of infected areas, and an export ban of ninety days on poultry from infected areas. There were high levels of compliance. Bird owners agreed to surrender birds for compensation, and that compensation was among the highest in the world for such programs. Firms were paid 80 percent to 100 percent of market value of the birds. Education efforts were taken to encourage compliance and safe handling techniques.[18]

In the next phase, longer-term planning and implementation was undertaken to improve public health measures and to win back export markets. The government stockpiled antivirals and developed quarantine protocols. Industrial interests and the interests of small farmers clashed over the use of vaccines. Small farmers raising poultry for local markets supported the idea of mandating vaccines. Industrial producers were vehemently opposed to this because of the cost and competitive issues and, more importantly, because trade partners in Europe and Japan would not buy vaccinated poultry. Not surprisingly, the government went to great lengths to protect and favor the industrial sector's interests over that of small- and medium-sized players. Vaccines were not used, but extensive culling and public education campaigns were conducted.[19]

Thaksin's clear political and economic connection to commercial and agribusiness interests involved in industrial poultry production demonstrates the overlapping social, political, and economic factors at work here. The perception inside and outside of Thailand has been that Thaksin's government responded openly and well to avian influenza; however, this is an incomplete picture. The government did not act quickly until large producers were impacted and motivated to accept government mandates and action. And there have been questions raised about how information was managed and manipulated about the extent of outbreaks and about vaccine use by different kinds of producers.[20] As far as demonstrating successful cooperation between national and supranational or international organizations, the Thai case demonstrates that countries can follow international recommendations and accept international funding, and yet retain a high degree of sovereign control. Rachel Safman in fact criticizes Thailand for marginalizing international veterinary health and public health experts who encouraged the use of vaccines in favor of prioritizing interests of strong domestic political interests, particularly from industrial poultry producers.[21] So, Thailand's success is attributable to following international recommendations, but why are they able to do so when other countries struggle at this? Political interests in combating the disease came from powerful stakeholders (poultry exporters), coupled with coherent political and policy action mandated from the center to rural areas; a strong political leader, and the ability to offer high levels of compensation to farmers for culling, allowed Thailand to produce successful results.

Handling disease outbreaks well—we see relative success in Vietnam and Thailand—is an example of synergistic fragmentation: different agencies, levels of government, and private actors have overlapping responsibilities where they need to coordinate on what they are doing, but where there are distinct objectives and roles for each actor.[22] When it does not work well, it exemplifies conflictive fragmentation and poor relations among actors, as was the case in Cambodia and Indonesia.

Vietnam

Vietnam was hit hard by early outbreaks of avian influenza. In 2004 there were twenty-nine human cases, and this jumped to sixty-one cases in 2005. Since then, Vietnam has had cases in only the single digits. How did Vietnam tackle this problem? Initially, Vietnam's government response was slow. Provincial officials were weak in identifying and reporting on the disease and the number of cases. When cases were reported, there was a significant lack of capacity to respond appropriately because of severe underresourcing of the national public health apparatus. As the number of cases increased in 2004, the government took more vigorous action. The prime minister and the Politburo of the Communist Party ordered swift action. There were efforts to securitize the outbreak. New language was used to frame the disease as a threat to security, such as "imminent danger" and "deadly threat," but these terms were used in an economic sense and not so much in a way that conveyed the threat was to life and health. There was a dramatic increase in the allocation of budget resources and much more cooperation with the WHO and the OIE. Vaccination efforts were stepped up: 212 million birds were inoculated. Vietnam wanted to avoid the mistakes that China made with SARS in hiding information and obstructing international efforts to understand and monitor the disease, so there were high levels of cooperation from Vietnam with the WHO and the OIE.[23]

The government of Vietnam was motivated to address this problem for both external and internal reasons. Externally, Vietnam did not want to scare off foreign investment and tourism. Thus, it wanted to at least appear to be able to deal with the disease outbreak in a competent manner, and to avoid suffering the criticism and outrage that China did for withholding information early on with the SARS outbreak. Internally, the regime's legitimacy is largely based on economic performance or success; communist ideology is no longer the motivating force or the legitimating ideology in Vietnam. Thus, the party maintains its hold on power (1) through its monopoly on controlling the institutions of power and most sources of information; and (2) by providing stability and economic progress. If either of these pillars appears to be shaky, the regime has significant cause for concern. "Security" in this instance was really the security of the regime, less than human security. So, Vietnam's success (similar to Thailand's) in addressing the avian influenza threat demonstrates its ability to provide public goods to its citizens, but less because the regime was concerned with human welfare and more because the regime was worried about economic damage and its own position of power.

Framing avian influenza as an economic threat made it easier to justify a culling program. There initially was some resistance to culling of sick or exposed flocks (although in Vietnam it is hard to actually know how much opposition there is), but the government's response was to enact tougher measures, not more lenient ones. By 2005, the government was taking the disease more seriously and it mounted a highly centralized and coordinated plan to address the problem. With help from a Canadian team of advisers, local groups were trained to recognize H5N1 and to take measures to curtail its spread. Information on the disease was spread throughout the country and more people were trained to deal with it.[24]

Some of the measures that Vietnam employed to combat avian influenza included mass poultry vaccination programs, culling, and prevention programs aimed at teaching people how to avoid exposure to the virus.[25] One of the things that made Vietnam's response so effective is the unitary nature of the state. The Communist Party controls all levels of government and once elites at the center (in the Politburo) became committed to addressing the problem, they were able to assert their policy preferences on local officials. Local-level officials have far less autonomy than in places like China or Indonesia, as shown below. Government control over the media and information dissemination also made it easier to impart one unified response or message on both local officials and the population more generally. Likewise, the comprehensive reach of the Communist Party, reaching down to the local village levels through farming cooperatives, women's unions, and so forth, made it more likely that Vietnamese farmers would comply with new controls than farmers in less authoritarian regimes like Indonesia. The power of the party over local governments and the realization of the (economic) security implications of failure help explain the synergistic and cooperative efforts in Vietnam to address HPAI.

Cambodia

Cambodia's response to H5N1 has been weak at best. Although the number of human cases reported in Cambodia in the initial period of Southeast Asia's outbreak was relatively small (only eight cases were identified between 2004 and 2008), despite enormous attention and funding from international donors Cambodia was not able to prevent a much larger outbreak from 2013 to 2014. In 2013 twenty-six human cases were identified, and then nine were reported in 2014. This increase in cases occurred after Cambodia had received $45 million in commitments from international donors to help prevent the spread of avian influenza.[26] As we also see below in the case of Indonesia, the two most significant problems that Cambodia faced in addressing the escalation of the number of human cases of H5N1 were (1) bureaucratic disagreements and posturing; and (2) weak infrastructure connecting national-level decisionmakers and policy ideas and the actions and priorities of officials at the local level.[27] Despite large sums of financial support and the knowledge of how neighboring countries were dealing with avian influenza, Cambodia's choices and strategies did not prevent a larger number of people from getting sick.

Small-scale farming is prevalent in Cambodia and 75 percent of households raise poultry.[28] These birds are for personal and local consumption. Cambodia is not a poultry exporter. Tourism is a significant contributor to the country's GDP, to the tune of about 14 percent annually, so there is a fear in Cambodia of the economic repercussions of a bird flu pandemic.[29] Despite this concern, the government has shown a lack of willingness to allocate domestic resources and use political capital to better address the health concerns (to animals and humans alike). As awareness and a sense of urgency about H5N1 built in 2003–2005, Cambodia attracted an enormous amount of assistance from the US Agency for International Development (USAID), the WHO, the UN Children's Fund (UNICEF), the FAO, the OIE, the EU, Japanese and Australian aid organizations, and dozens more. Sophal Ear reports the involvement of possibly more than 160 NGOs worldwide.[30] One example is of a program funded by USAID: a massive public health campaign was undertaken, along with creating a child's superhero action figure to promote it, to educate people about healthy behavior around poultry. This included advising people to fence in their birds, to quarantine new poultry, and to use protective equipment in dealing with sick, or possibly sick, birds.[31] Although public health campaigns geared toward changing people's behavior around birds is an important part of a country's response to avian influenza, it is not the only important factor. Some of these measures require not only changing attitudes and behaviors, but financial costs as well. Given the high poverty levels in Cambodia, it is difficult to see how these policies could work without direct financial support to those who need to change their behavior. Also important are culling measures, restrictions on moving poultry in or out of infected regions, and vaccination programs. Cambodia has not implemented these types of measures with any degree of success:

> The avian influenza response in Cambodia has been dominated by external, donor-led efforts. A huge range of actors exist, yet the state provides a barrier between external aid funding and local level action on the ground. The siphoning of aid flows to fuel patronage networks is a well known phenomenon in Cambodia, and the avian influenza response has added to this dynamic. Well connected officials, linked to political networks and the ruling party, are able to benefit, with aid efforts directed to certain areas and activities.[32]

There has been no official policy mandating culling of birds or offering compensation to farmers for doing so. Explanations for the lack of these policies, according to Ear, are that officials feared that levels of compliance would be very low, and that the incentives to cheat the system would be very high, along with reluctance of the government to use either state revenue or donor funds for these purposes.[33] Instead, it seems that officials wanted to use donor funds for programs that would benefit the officials themselves, such as public health campaigns that they could oversee or money for data collection, rather than channeling the funds to people and communities directly involved with changing poultry-raising practices. Again, we see that rather than setting up better procedures and mechanisms for providing public goods, government officials were using resources for private gain.

In Cambodia's capital, Phnom Penh, the two government ministries most directly involved in addressing and combating H5N1 are the Ministry of Health (MOH) and the Ministry of Agriculture, Forestry and Fisheries (MAFF). Work by these agencies has been remarkable not for their successes and capacity building from the influx of aid money, but rather for finger-pointing and blame. Finger-pointing ensued over which agency should have been the first to identify and monitor outbreaks among humans (MOH's responsibility) or in birds (MAFF's turf), and there was competition for funds and attention from foreign donor groups. Even if these agencies could work better with each other, the other part of the problem has to do with governance infrastructure. In Cambodia, the power of the ruling party, the Cambodian People's Party (CPP), comes from an extensive support base in the countryside. The party is able to exert influence through a dense network of village, district, and commune chiefs. Their loyalty is often to military officials and political leaders who can buy their patronage.[34] Thus, even if the central bureaucratic agencies did work together to implement a wide variety of programs, they would still face an uphill battle to get local compliance. Compound this with weak physical infrastructure of things like roads, schools, and health clinics, and it becomes a mammoth task to think about implementing a comprehensive health initiative in Cambodia even with foreign aid money pouring into the country. Lack of compensation for culling flocks, bureaucratic infighting, poor infrastructure, and repurposing of aid money resulted in the poor responses to HPAI in Cambodia. Here, fragmentation was full of conflict and competition, and the outcome was an inability to effectively combat the disease.

Indonesia

Indonesia, like most of Southeast Asia, is home to millions of people who raise a small number of birds for personal consumption. Bird markets are quite common and people live in close proximity to their animals. Although it has manifested differently, like Cambodia, Indonesia has tried to leverage the threat of an avian influenza pandemic in its interactions with international agencies to change domestic political dynamics.

H5N1 is widespread in Indonesian poultry, including both wild and domestic birds. Early on, Indonesia engaged in misleading reporting and denials about human transmission. Until January 2004, the Indonesian agriculture minister declared that H5N1 was unable to be transmitted to humans and that only visibly infected poultry needed to be culled,[35] neither of which is correct. Resource allocation to address avian influenza was also quite low and there was little institutional support at the local or national level for greater measures. Paul Forster's study on avian influenza in Indonesia found weak policy implementation in local areas and only small amounts of money being used to support officials' efforts. Local district officials selectively reinforced or contradicted policies enacted by Jakarta.[36]

There are several obstacles to a better response to H5N1 in Indonesia, including unclear and fragmented policy implementation in local areas. Since the end of Suharto's rule in 1998 and the development of democracy, policymaking and implementation have changed drastically. To understand Indonesia's efforts to combat avian influenza, the most significant political change has been the decentralization of decisionmaking: no longer a unitary state, policymakers in Jakarta must now work with, and compete and contend with, powerful local actors.

In 2005, new guidelines, the International Health Regulations, were promulgated by the WHO to deal with infectious disease outbreaks (partly these regulations came about because of the global experience with SARS). The International Health Regulations reflected greater development of reporting requirements, information sharing, and coordination among key actors. These are the elements that make up global health governance. Prior to the 2003 SARS outbreak, the WHO was only playing a role in monitoring diseases and doing border surveillance. After SARS, there was greater discussion and focus on the quality of domestic health systems (particularly of developing countries) and how well these systems could manage EIDs for the broader global benefit. The 2005 International Health Regulations allowed the WHO to declare a public health emergency of international concern (PHEIC) for any disease. This allows the WHO to collect data from the field. Concern over H5N1 resulted in $2 billion in pledges by the United States, Japan, the EU, and Australia to spend in countries hardest hit by the disease. Indonesia was earmarked to receive $130 million. And additional money came to Indonesia from NGOs already working in the country on public health programs.[37]

A great deal of international attention suddenly turned to Indonesia in 2006 when Minister of Health Supari announced the decision to stop sharing samples of the H5N1 virus with the WHO. And in 2008, Triono

Soendoro, director of the Health Research and Development Agency of the Health Ministry, announced that Indonesia was shutting down NAMRU-2 in Jakarta. "The NAMRU-2 Detachment was established in Indonesia in 1970 and is one of the most advanced facilities of its kind in Southeast Asia. Its work was done in cooperation with and under the auspices of the Indonesian Health Research and Development Agency."[38] The explanation offered was that it was being closed down over complaints about "viral sovereignty" and accusations that the lab was using information and samples from Indonesia to manufacture vaccines that were sold and distributed elsewhere for profit. Indonesia was concerned that it was not sharing in the economic benefits from this. Supari framed her decision at the time as one of inequality between rich nations and poor, North and South; Indonesia was standing up to a global community that favored profits for big pharmaceutical companies over the needs of a developing country.[39]

While Supari's action plays into discussions about Indonesia's relationship with the global community, there really were two audiences and political explanations for her decision. First, it did help stir up nationalism in the country at a time when commodity prices were high and there were economic concerns within the country, thus the appeal of populist anti-imperial rhetoric and action. But perhaps more importantly, attacking NAMRU-2 demonstrated an internal power struggle between the Ministry of Health (a relatively weak agency) and other ministries such as the Ministry of Forestry, Ministry of Agriculture (MOA), and Ministry of Industry as well as between the MOH and powerful interests of agribusiness and livestock industries. Shahar Hameiri argues that the MOH was not powerful enough to take on these well-funded and well-connected interests and actors within Indonesia, but found a way to assert their interests in going after the large number of international agencies cooperating in Indonesia on disease outbreaks.[40] These international organizations included the WHO, the FAO, the OIE, and the World Bank, to name a few. Their work in Indonesia involved coordinating with numerous government and nongovernment agencies to combat infectious diseases. The Ministry of Health over time became just one of a handful of ministries involved in HPAI. Managing infectious diseases was no longer the exclusive purview of the Directorate General of Disease Control within the MOH; other ministries also had a hand in addressing H5N1, as did various committees set up specifically to deal with HPAI. The proliferation of international organizations and bureaucratic agencies dealing with some aspect of disease outbreaks indicates the high level of complexity involved, and thus more opportunities for problems and competition; this is exactly what can be seen in Indonesia.

Although much has been made of the closing of NAMRU-2 as an indication of Indonesia's unwillingness to work with the global community on H5N1, Indonesia did not stop sharing samples of H5N1 from poultry with the FAO, and other kinds of cooperation with the WHO persisted even while NAMRU-2 was being targeted and finally shuttered in 2010. Indonesia's decision did disrupt the work being done in the country and the region by the WHO and the Global Influenza Surveillance Network (GISN) at a time when fear of H5N1 was quite high. This led to a new agreement by the World Health Assembly (WHA) in 2011 on pandemic influenza preparedness, which was the culmination of four years of negotiation among the WHO's member states.

The framework of the WHA agreement improves influenza virus sharing and access to vaccines and other benefits. Member states agreed that the agreement lays the groundwork for better preparedness and better access to tools and knowledge.[41] Concern and attention by international organizations presented internal actors like Minister Supari with the opportunity to assert independence and raise the stakes of what Indonesia could get from the international system. Somewhat ironically, the Ministry of Health was better able to assert power over the WHO than over internal interests inside of Indonesia. In shutting down NAMRU-2, the MOH was able to redirect some of the funding that went to NAMRU-2 to other labs under the auspices of the ministry, and Indonesia did receive more vaccines and antivirals than it had been getting previously.[42] Indonesia was also able to bolster its partnerships with labs from Japan and elsewhere to fill some of the gaps left by shutting down NAMRU-2, and the dispute over virus sharing may have helped attract funding from within government coffers in Indonesia and from international donors to purchase diagnostic equipment for other labs and to increase vaccine production.[43] Both the case of China's withholding of information and samples during the SARS outbreak and Indonesia's decision to stop sending the WHO HPAI samples signal the challenges that the WHO faces in getting countries to comply with international health governance.

Some of the measures that Indonesia should undertake to address HPAI include the following:

- develop a certification program for poultry production;
- develop a national veterinary service to build capacity of animal husbandry services;
- develop better relations and partnerships between government agencies and poultry producers to monitor poultry production and standards as well as veterinary care;
- improve knowledge of basic disease transmission;

- develop a centralized strategy to monitor viruses and vaccinations; and
- bolster farmer incentives to use veterinary services and compliance with biosafety measures.[44]

Indonesia is doing some of these things, but not in a uniform or consistent basis across the country. There are many ongoing projects in the country run by NGOs (in partnership with national and local government agencies) to address gaps in Indonesia's response to HPAI. For example, the Community-Based Avian Influenza Control Project (CBAIC) is a collaboration between USAID and the FAO, the WHO, the International Livestock Research Institute, the Ministry of Agriculture Avian Influenza Campaign Management Unit, the Ministry of Health, and local government officials. CBAIC began its work in Indonesia in 2006 to extend surveillance efforts by educating communities to monitor animals and humans at risk for HPAI. CBAIC designs and implements community mobilization programs in West Java. It has developed mass media campaigns to change animal husbandry practices to minimize the risk of HPAI. It is working to find a commercial poultry partner to develop and test biosecurity advisory service models.

At the local level, CBAIC has been working with the Indonesian Red Cross and the organization Muhammadiyah to train village and district coordinators. As a result, 80 master trainers were certified, who trained 1,753 volunteer subdistrict avian influenza coordinators and 25,210 village avian influenza coordinators in 124 districts across 9 provinces in western Indonesia. These trainers play a key role in promoting and disseminating information about minimizing the risk of avian influenza transmission between animals and humans. CBAIC also designed and implemented materials to be distributed and used in communities to promote behavior modification (educational materials on improving hygiene and sanitation to reduce disease risk) and reporting techniques to support surveillance and response programs. By the end of 2009, these programs numbered 11,270 across 1,000 villages.[45]

From 2011 to 2013, Development Alternatives, Inc. (DAI) also helped run, as part of USAID's Indonesian Avian and Pandemic Influenza Program, the Strategies Against Flu Emergence (SAFE). The goal of the program is to assist the government and the private sector to strengthen prevention and response to HPAI. This program builds on the success of programs by the MOA and MOH, local governments, and the CBAIC and international partners. The focus of these programs is twelve high-risk districts in West Java and Banten.[46]

While it is hard to know how effective such measures are (rates of human HPAI cases have fallen dramatically since 2011), DAI claims that residents in CBAIC-mobilized communities are more likely than those in villages without these programs to practice safer poultry-raising and processing behaviors, and it is reported that people engage in better handwashing and cleaning of clothes, tools, and surfaces that have been in contact with poultry. Likewise, reporting is better in these villages than elsewhere; poultry deaths are more likely to be reported to local leaders or animal health officials than in communities not mobilized by CBAIC.[47] The wider sense of actions at the local level to address HPAI is less optimistic. Isabel Gertler finds that, with decentralization, public health programs vary tremendously across the country. It is challenging to implement uniform programs and defense measures against HPAI and to spread information about best practices, but that does not mean that there are not some successes. Gertler finds that there is weak compliance with culling and vaccination programs because there are few incentives for people to participate or comply, and there is little funding to compensate farmers for culling birds. There is a lack of trained veterinarians to do proper monitoring or inspecting.[48] Thus, in Indonesia there are extreme examples of fragmentation and complexity in governance architecture. The overall effect of this is that Indonesia has not been as effective in combating the disease as Thailand and Vietnam. It is possible that the fragmentation could be workable in some instances, but the overall picture does not provide evidence of this.

Much of the domestic and international fear and attention over HPAI has dissipated. With overall transmission rates lower, and stockpiles of vaccines at a comfortable level, it is tempting to become complacent about the disease. This would be a mistake. Most of the conditions that made the region susceptible to the threat of a pandemic outbreak still exist, and it is not at all clear if Indonesia and Cambodia have adequately implemented best practices so that they would be able to quickly and effectively deal with an outbreak. In April 2015, Indonesia announced that it would resume giving the WHO access to viral samples. Because of Indonesia's earlier decision to withhold virus samples from the WHO, a new resolution has been put in place whereby pharmaceutical companies who want access to virus samples are required to choose to either donate a sum of money to the donor country for research, give up the licensing so local manufacturers can mass produce the vaccine themselves, or reserve a certain percentage of their production for the WHO to access during a pandemic. Head of the research unit at the Indonesian Ministry of Health, Tjandra Yoga Aditama, announced that Indonesia would fully cooperate and encouraged the WHO to bring on board as many vaccine makers as possible, including Indonesian vaccine makers such as Bio Farma and Kimia Farma.[49]

Indonesia's initially defensive and weak response to HPAI stemmed from bureaucratic jockeying, low levels of compliance at the local levels, and poor relations between Jakarta ministries and global organizations. Over time NGOs, working with government agencies, were able to improve preventive measures and bring infection rates down. Indonesia's responses, and infection rates over the past five years, have been better than Cambodia's, but not as effective as Vietnam's or Thailand's. Greater power of local officials and low levels of compensation for culling, along with policy resistance by agribusinesses, have made it more difficult to achieve better outcomes in Indonesia.

International Organizations and Global Health Governance

Outbreaks of avian influenza occurred in the midst of a period of increased attention to global health governance. As a result of increasing awareness of the occurrence of new and deadly pathogens, such as SARS, governance efforts have increased over the past twenty years. New organizations and regimes have been developed to monitor and respond to public health problems, there has been unprecedented growth in funding for global health, and there has been increased influence from policymakers, activists, and donors (and NGOs) who view global health as a critical international issue. The magnitude of the threat prompted significant levels of funding and increased attention to global health governance. Global health governance in addressing HPAI, as described below, includes new organizations, programs, and legal mechanisms as well as new cooperation among states, local actors, international institutions, and NGOs. This is what scholars of networked and fragmented governance are describing with the proliferation and increased complexity of organizations, and levels of government and nongovernment actors involved in issues such as the spread of new and deadly diseases.[50] Despite the increased attention and funding, and partnerships among government agencies, private forces, and international organizations, and despite efforts to bring in a variety of stakeholders, there remain significant gaps in disease responses to global health threats.[51] A number of international organizations and agencies are at the forefront of dealing with HPAI: the WHO, the FAO, the OIE, and the CDC are just the most significant ones. The WHO, the FAO, and the OIE are the primary institutions that deal with avian influenza. The WHO has developed and coordinates the Global Influenza Surveillance Network, previously entitled the Global Influenza Surveillance and Response System (GISRS), which are laboratories in developed countries that have the responsibility for vaccine development and testing for HPAI. GISN relies on cooperation from member states and also serves as a global alert mechanism for the emergence of influenza viruses with pandemic potential.[52] Over the past twenty years, there has been a shift in the mission and activities of global health international organizations. Once these institutions were tasked with watching borders for disease transmission and reporting on findings and statistics but, with the rise of new and more harmful infectious diseases (e.g., SARS, Ebola, and Zika), there has been a shift in what these international organizations do. This is the securitization process that many have written about.[53] Reframing disease outbreaks as security threats means greater attention, greater funding, and greater capacity building at both the state level and the international level. The magnitude of the threat in the case of HPAI is what made securitization possible. Although the number of cases of avian influenza has been relatively low, the potential for it to become a pandemic has caused considerable concern and thus the increased funding and shift in language about how we think about diseases as a security threat.

The WHA is the decisionmaking body of the WHO. Delegates from all member states meet in annual assemblies in Geneva to determine policies for the WHO.[54] Prior to the 2003 SARS outbreak, the WHO was responsible for monitoring six diseases and carried out border surveillance of these diseases. The WHO had generally operated as a scientifically grounded, technically focused institution. Instead of binding treaties and international law, the WHO mainly operated by developing norms and expectations of behavior.[55] Global concern over SARS, and the poor handling of it in China, led the WHO to increase its attention to domestic health systems, and there were significant efforts to ramp up its ability to manage EIDs to mitigate, slow, or prevent broader global spreading of the diseases. The WHO is able to declare a PHEIC for any disease and this allows it to collect data directly from the field.[56] In 2000, the Global Outbreak Alert and Response Network (GOARN) was created as a system of technical institutions to coordinate responses to disease outbreaks.

The network is comprised of technical institutes, research facilities, universities, health organizations, and other technical networks. The WHO coordinates this network. GOARN has been deployed in the context of HPAI in 2004 in Cambodia, Indonesia, Thailand, and Vietnam. Health officials within research facilities, labs, and hospitals work to support national health authorities who are called on by the WHO to provide rapid reporting and support for outbreak response mechanisms. The goal is to influence the development of local, national, and international capabilities and longer-term partnerships and initiatives to combat disease. This is an example of how international organizations attempt to provide public goods. By monitoring and addressing disease outbreaks, the goal is to keep the outbreaks from spreading and to assist individual countries in being able to cope with them. And by providing states with the assistance to comply with best practices, the WHO is giving countries incentives (in the form of information, tools, and resources) to cooperate.

Some of the activities spearheaded by GOARN include gathering data, lab support, risk assessment, logistics, and posting information to a SharePoint website to facilitate communication within and across health networks. GOARN was developed in response to SARS and seems tailor-made for EIDs like H5N1.[57] In addition, GOARN helps oversee local efforts by government and NGO agencies to develop training materials, field team leadership, disease-specific responses, and rapid containment operations if necessary. The goal is to develop regional capacity for international responses. International Health Regulations (IHR) are legally binding agreements drafted by the WHO and agreed on by member states. The new IHR in 2005 had a broader purpose of preventing and protecting from failures of public health systems to respond adequately to the international spread of diseases. The goal is to minimize public health risks and avoid disruptions to international travel and trade.[58] New regulations require states to develop specific minimum core public health capacities for surveillance and response activities and reaffirm the central role of the WHO in assisting countries in controlling the international spread of infectious diseases. States had until 2012 to meet their IHR core surveillance and response requirements, particularly those measures to be put into place at airports, ports, and certain land borders. A majority of states, however, requested and were granted a two-year, or in some cases four-year, extension to the deadline.[59] There are few enforcement mechanisms in place other than the ability to withhold funds or to internationally shame countries for noncompliance. Neither of these measures seem likely to be chosen since the WHO has a vested interest in getting countries to comply eventually, rather than shutting them out because of censure. GOARN would undertake to provide guidance and assistance to mobilize international teams of experts for on-site assistance. The new IHR have increased the awareness of countries like Indonesia about sharing information on acute public health events.[60] This is an example where countries would welcome the benefits of global assistance in coping with disease outbreaks, but, when it comes down to it, they often are prone to free-riding (not supplying the requested information, or not bolstering domestic public health systems), instead of complying with what's asked of them. Despite these tendencies, the new requirements and protocols do seem to be moving countries slowly toward this collective and supranational behavior.

In Indonesia, the WHO oversaw programs directly: the participatory disease surveillance and response (PDSR) and district surveillance officer (DSO), which were run through the Jakarta offices of the WHO and the FAO. Even while NAMRU-2 was being shut down, Indonesia did share data with the WHO on H5N1, and it shared virus samples with the FAO.

The FAO leads the world in addressing issues of world hunger and the global food supply. It works closely with the OIE to combat HPAI. Like their work on other infectious diseases, the FAO and the OIE aim to develop regional capabilities to fight against transboundary animal diseases and to assist in capacity building within countries to monitor and control disease outbreaks. Their joint efforts aim to address the complexity of problems that impact market access, food supply chains, and human welfare, all of which are hurt by H5N1. The FAO and the OIE aim to improve both human and animal health by focusing efforts on developing regionally led efforts with the participation and agreement of relevant stakeholders, developing regional and global early warning systems, and promoting greater research on strategic disease management and control.[61] The FAO and the OIE have also done a great deal of work to promote awareness of biosecurity and to work with affected countries to prevent incursion and spread of disease. Such measures include segregation and decontamination measures to help contain the spread of HPAI. Specific practices vary somewhat for larger commercial producers and smaller backyard flocks, and the FAO and the OIE have been instrumental in drafting guidelines to address the needs of different sectors in the market and throughout the production-consumption chain.[62]

ASEAN is a logical place to look for coordination and cooperation on HPAI. Yet the organization has played a relatively minor role in combating the spread of the disease. Two ministerial-level ASEAN bodies, ASEAN Ministers on Agriculture and Forestry and ASEAN Health Ministers Meeting (AHMM), have discussed and provided direction to working groups and experts under ministerial control. For example, working groups and task forces have been created to take part in addressing HPAI: the ASEAN Sectoral Working Group on Livestock (ASWGL) and the ASEAN Highly Pathogenic Avian Influenza Task Force from the animal health sector and the Senior Officials Meeting on Health Development (SOMHD) and ASEAN Expert Group on Communicable Diseases (AEGCD) from the public health sector have met to formulate coordinated multiagency and multisectoral approaches to preventing, controlling, and eradicating HPAI in the region. These agencies have been asked to work with the international organizations and institutions mentioned earlier to execute policies and practices to help report, monitor, and combat the disease. To do this, ASEAN has tried to improve institutional capacity to coordinate public health systems and responses. Different countries were made responsible for overseeing different parts of this task. For example, Malaysia was tasked with overseeing improvement of capacity of national and regional laboratories in diagnostics, surveillance, and rapid response; and Thailand was charged with coordinating improvement of national and regional capacity in epidemiological surveillance, preparedness, early warning of outbreaks, and rapid response to emerging infections. Finally, Indonesia has coordinated improvement of capacity of the ASEAN Disease Surveillance Network to meet the needs of ASEAN member countries in emerging infectious disease surveillance, preparedness, and response.[63] ASEAN has served as a vector for coordination with other countries and international organizations. It has worked with the Asian Development Bank and with the FAO and the WHO to implement a grant to strengthen early detection, reporting, and control of avian flu outbreaks and to assist with human cases of the disease. ASEAN also received funding from Japan and Australia to bolster initiatives within the region. In 2006, Japan authorized $70 million to help ASEAN integration with other international organizations and among member states to fight HPAI. The assistance included a stockpile of 500,000 courses of Tamiflu and supplies such as test kits, antiseptic solutions, masks, rubber gloves, gowns for 700,000 persons.[64] Thus, Southeast Asia states have incentives to comply with best practices when wealthier countries are helping to subsidize the costs.

Several developed nations have been significantly involved in efforts to monitor and combat HPAI. The United States offered financial support and technical assistance through the CDC, USAID, and dozens of NGOs. In 2005, the United States provided $31.3 million in support to international agencies and recipient countries, and this increased to $280 million in 2006 and $215 million in 2007. Germany, Japan, and Australia have also devoted significant resources to dealing with HPAI. Thailand, after having success in combating its own outbreak, proposed creating an ASEAN Animal Hygienic Fund and pledged $300,000 to start the program.[65] The program would help with border surveillance and control measures as well as encourage better information sharing within ASEAN. Asia Pacific Economic Cooperation, APEC, the large Pacific basin organization, endorsed the US-led International Partnership on Avian and Pandemic Influenza (IPAPI) initiative. This comprised a series of steps on preparing for and mitigating an HPAI pandemic, and it called for collective and transparent measures to exchange information and expertise. In 2006, APEC ministers adopted a five-part Action Plan on Prevention and Response to Avian and Influenza.

Conclusion

Southeast Asian countries' responses to HPAI have varied. Vietnam's and Thailand's experiences demonstrate that developing countries can deal effectively with an EID and that complex fragmented governance mechanisms can work, while Cambodia and Indonesia provide examples of the challenges and obstacles to achieving more comprehensive success. Vietnam's and Thailand's ability to lower the number of people infected with H5N1 illustrate the importance of government action and its ability to implement policy choices at the local level. The choice and budget allocations to compensate farmers for culling sick or exposed flocks, and the use of enforcement measures to make sure compliance happens, are critical to weakening and halting transmission of the disease. The fact that Vietnam is a unitary and one-party state where party networks extend into all economic sectors and down to the local levels in rural areas, plus the nature of interest group preference of poultry exporters in Thailand in supporting strong government action, makes it easier for Vietnam and Thailand to follow best practices at disease control and prevention.

Less successful outcomes in Cambodia and Indonesia are attributable to greater divisions and sources of competition and conflict between national-level bureaucrats and local officials, which has led to less (or no) money for culling birds, weaker enforcement of policy decisions, and less coherent or uniform implementation of best practices throughout the two countries. Although Cambodia and Indonesia have received millions of dollars in financial assistance, money alone is not enough to overcome the bureaucratic squabbling, corruption, conflicting interests of donor organizations and government agencies, and poor governance infrastructure in each country. These are examples of fragmented governance going awry.

As Shahar Hameiri and Lee Jones discuss in their book *Governing Borderless Threats,* nontraditional security issues are fundamentally problems of governance. Their work shows how local sociopolitical conflicts shape efforts and outcomes to deal with problems such as climate change, pandemics, and other nontraditional security threats. Hameiri and Jones find wide variation across countries and even within a country in how well or how poorly disease outbreaks are handled. Such differences reflect different political circumstances and dynamics.[66] Regional and international organizations, along with wealthy countries such as the United States and Japan and international NGOs, have stepped up funding and supranational efforts to monitor, study, and control EIDs. Yet as we illustrated in this chapter, these measures can be effective only if national and local levels of government, and important economic actors, share the same priorities and are willing and able to implement recommendations and protocols. International organizations can help build domestic capacity, but states must see it as in their interests and must have the political will to improve or overcome poor governance abilities. This applies to states' abilities to provide public goods to citizens more generally. Addressing disease outbreaks is an essential job of government to keep citizens healthy and safe. However, as we show repeatedly in this book, providing this essential element of human security is challenging. In attempting to provide public goods, governments are stymied by competing political interests and by governance weaknesses. While HPAI seems mostly under control in Southeast Asia today, it remains to be seen if the lessons learned and the health measures developed to address H5N1 will help prevent the next pandemic.

Notes

1. Melissa G. Curley and Jonathan Herington, "The Securitisation of Avian Influenza: International Discourses and Domestic Politics in Asia," *Review of International Studies* 37, no. 1 (2011): 141–166.

2. US Centers for Disease Control (CDC), "Why Global Health Security Matters," February 13, 2014, www.cdc.gov/globalhealth/security/why.htm.

3. Severe acute respiratory syndrome (SARS) is a serious virus. Between November 2002 and July 2003, an outbreak of SARS in southern China eventually caused more than 8,000 cases and 774 deaths in 37 countries, with the majority of cases in Hong Kong. China's initial response to SARS was to withhold information from international bodies and the Chinese public, and it took months before it agreed to cooperate on sharing information and to be forthcoming about outbreaks. No cases of SARS have been reported worldwide since 2004.

4. Curley and Herington, "The Securitisation of Avian Influenza," 151–153.

5. Shahar Hameiri and Lee Jones, *Governing Borderless Threats: Non-Traditional Security and the Politics of State Transformation* (Cambridge: Cambridge University Press, 2015), Introduction and chap. 4.

6. CDC, "Highly Pathogenic Asian Avian Influenza A (H5N1) in People," June 11, 2015, www.cdc.gov/flu/avianflu/h5n1-people.htm.

7. World Health Organization (WHO), "Avian Influenza Fact Sheet," updated November 2016, www.who.int/mediacentre/factsheets/avian_influenza/en/.

8. The CDC states that it is possible that very close contact with an infected person may transmit the disease from person to person. "Currently, HPAI Asian-origin H5N1 virus does not transmit efficiently from person to person. Some cases of limited, non-sustained human-to-human transmission have likely occurred." Such transmission has occurred only through prolonged and close contact within families. CDC, "Highly Pathogenic Asian Avian Influenza A (H5N1) in People."

9. Isabel Gertler, "A Focus on Factors Affecting Avian Influenza in Indonesia." (Sackville, New Brunswick, Canada: Atlantic International Studies Organization, December 2009), 1–9.

10. Emma Chanlett Avery, *International Efforts to Control the Spread of the Avian Influenza (H5N1) Virus: Affected Countries' Responses,* Report for Congress (Washington, DC: Congressional Research Service, August 24, 2006).

11. Ibid., 2–7.
12. "Indonesia Bans Navy Medical Research Unit," *Navy Times,* April 10, 2008; Shahar Hameiri, "Avian Influenza, 'Viral Sovereignty,' and the Politics of Healthy Security in Indonesia," *Pacific Review* 27, no. 3 (2014): 333–356.
13. Supamit Chunsuttiwat, "Response to Avian Influenza and Preparedness for Pandemic Influenza: Thailand's Experience," *Respirology* 13, no. 1 (March 2008): S36–S40.
14. For more on this period in Thailand's political economy see Amy L. Freedman, *Political Change and Consolidation: Democracy's Rocky Road in Thailand, Indonesia, South Korea and Malaysia* (New York: Palgrave Macmillan, 2006).
15. Rachel M. Safman, "The Political Economy of Avian Influenza in Thailand," Working Paper No. 18 (Brighton: Social, Technological and Environmental Pathways to Sustainability [STEPS] Centre, 2009).
16. Cockfighting is also prevalent in Thailand and the risks of avian influenza are quite high in the raising, training, and participation in this spectacle.
17. Safman, "The Political Economy of Avian Influenza in Thailand," 14.
18. Ibid., 15–16.
19. Thanawat Tiensin, P. Chaitaweesub, T. Songserm, A. Chaisingh, W. Hoonsuwan, C. Buranathai, et al., "Highly Pathogenic Avian Influenza H5N1," *Emerging Infectious Diseases* 11, no. 11 (2004): 1664–1672.
20. Safman, "The Political Economy of Avian Influenza in Thailand," 51.
21. Ibid., 52.
22. Frank Biermann, Philipp Pattberg, Harro van Asselt, and Fariborz Zelli, "The Fragmentation of Global Governance Architectures: A Framework for Analysis," Global Environmental Politics 9, no. 4 (2009): 14–40.
23. Curley and Herington, "The Securitisation of Avian Influenza," 151–153.
24. Ibid., 155; Gertler, "A Focus on Factors Affecting Avian Influenza in Indonesia," 1–9.
25. Avery, *International Efforts to Control the Spread of the Avian Influenza (H5N1) Virus.*
26. Sophal Ear, "Avian Influenza: The Political Economy of Disease Control in Cambodia," *Politics Life Science* 30, no. 2 (2011): 2–19.
27. Curley and Herington, "The Securitisation of Avian Influenza," 163.
28. FAO, "EMPRES Transboundary Animal Diseases Bulletin: Special Issue Avian Influenza," 2007, ftp://ftp.fao.org/docrep/fao/007/y5537e/y5537e02.pdf, 18–39.
29. Sophal Ear, "Cambodia's Patient Zero: The Political Economy of Foreign Aid and Avian Influenza," Working Paper No. 398 (Stanford: Stanford Center for International Development, September 2009).
30. Ibid.
31. Ibid., 11.
32. Ibid., 20.
33. Ibid.
34. Ibid., 6–9.
35. Curley and Herington, "The Securitisation of Avian Influenza," 156–157.
36. Paul Forster, *The Political Economy of Avian Influenza in Indonesia* Pathways to Sustainability Book Series (Sussex: Institute of Development Studies, STEPS Centre, March 10, 2009).
37. Hameiri, "Avian Influenza, 'Viral Sovereignty' and the Politics of Health Security in Indonesia," 338–340.
38. "Indonesia Bans Navy Research Unit."
39. Curley and Herington, "The Securitisation of Avian Influenza," 158.
40. Hameiri, "Avian Influenza, 'Viral Sovereignty' and the Politics of Health Security in Indonesia."

41. WHO, "Sixty-fourth World Health Assembly Closes After Passing Multiple Resolutions" (Geneva: WHO, May 24, 2011).

42. Although Siti Fadilah Supari's decision to stop sharing samples with the WHO and her shuttering of Naval Medical Research Unit No. 2 (NAMRU-2) were seen as highly provocative and political acts, it is interesting to note that her successor as minister, Endang Rahayu Sedyaningsih, a well-respected former researcher at NAMRU-2, did not choose to reinvite NAMRU-2 to return to Indonesia. Shahar Hameiri argues that this signals that the closure of the facility was more about internal political struggles among ministries than about effective combating of disease. Hameiri, "Avian Influenza, 'Viral Sovereignty' and the Politics of Health Security in Indonesia."

43. Hameiri, "Avian Influenza, 'Viral Sovereignty' and the Politics of Health Security in Indonesia," 20.

44. Food and Agriculture Organization (FAO), "Approaches to Controlling, Preventing and Eliminating H5N1 Highly Pathogenic Avian Influenza in Endemic Countries" (Rome: FAO, 2011).

45. Development Alternatives, Inc. (DAI), "Indonesia—Community-Based Avian Influenza Control (CBAIC) Project," accessed October 24, 2017, https://www.dai.com/our-work/projects/indonesia-community-based-avian-influenza-control-cbaic-project.

46. US Agency for International Development (USAID) and US Department of State, "Strategies Against Flu Emergence (SAFE) Project (2011–2013)—Ongoing," www.k4health.org/toolkits.

47. DAI, "DAI-Led Project Credited with Reducing Risk of Avian Influenza in Indonesia," May 10, 2010.

48. Gertler, "A Focus on Factors Affecting Avian Influenza in Indonesia," 1–9.

49. Dessy Sagita, "Indonesia's Health Ministry Gives WHO Access to Bird Flu Samples Again," *Jakarta Globe,* April 27, 2015.

50. Jessica R. Daniel, Sandra Lee Pinel, and Jase Brooks, "Overcoming Barriers to Collaborate Transboundary Water Governance: Identifying Local Strategies in a Fragmented Governance Setting in the United States," *Mountain Research and Development* 33, no. 3 (August 2013): 215–224.

51. David P. Fidler, "The Challenges of Global Health Governance," Working Paper (New York: Council on Foreign Relations, May 2010).

52. WHO, "Global Influenza Surveillance and Response System (GISRS)," updated September 2017, http://www.who.int/influenza/gisrs_laboratory/en.

53. Curley and Herington, "The Securitisation of Avian Influenza," 149–150.

54. WHO, "World Health Assembly," updated August 2017, www.who.int/mediacentre/events/governance/wha/en/.

55. Fidler, "The Challenges of Global Heath Governance," 4.

56. Hameiri, "Avian Influenza, 'Viral Sovereignty' and the Politics of Health Security in Indonesia," 6. It is worth noting here that states must allow the WHO entry and activity in their country.

57. John S. Mackenzie, Patrick Drury, Ray R. Arthur, Michael J. Ryan, Thomas Grein, Raphael Slattery, Sameera Suri, Christine Tiffany Domingo, and Armand Bejtullahu, "The Global Outbreak Alert and Response Network," *Global Public Health* 9, no. 9 (2014): 1023–1039.

58. Ibid., 1035.

59. WHO, "About IHR," www.who.int/ihr/about/en.

60. Mackenzie et al., "The Global Outbreak Alert and Response Network," 1035.

61. World Organisation for Animal Health (OIE), "The Global Framework for the Progressive Control of Transboundary Animal Diseases (TADs)" May 24, 2004,

http://www.oie.int/fileadmin/Home/eng/About_us/docs/pdf/GF-TADs_approved_version24May2004.pdf.

62. "The Importance of Biosecurity in Reducing HPAI Risk on Farms and in Markets," paper prepared for the FAO International Ministerial Conference on Avian and Pandemic Influenza, New Delhi, December 2007.

63. ASEAN Secretariat, "ASEAN Response to Combat Avian Influenza" (Jakarta: ASEAN Secretariat, April 2006).

64. Ibid.

65. Avery, *International Efforts to Control the Spread of the Avian Influenza (H5N1) Virus,* 18.

66. Hameiri and Jones, *Governing Borderless Threats,* Introduction and chap. 4.

6

Migration: Divergent Interests of Sending and Receiving States

The movement of people across Southeast Asian borders is a significant transnational issue that holds the potential to create both conflict and cooperation. According to the International Organization for Migration, *migration* is the movement of people across borders or within a state and can encompass refugees, economic migrants, displaced peoples, and people moving for other purposes including family reunification.[1] Because this book is about transnational issues, we focus on people who cross state borders instead of on internally displaced persons (IDPs) or those who leave their place of habitual residence but remain within the same state.

Southeast Asia has experienced significant waves of migration produced by different political, social, and economic factors. Famines in China have sent exoduses of Chinese to Southeast Asia. During the colonial era, the British imported large numbers of Chinese and Indian workers to their colonies in Burma, Malaysia, and Singapore. Following independence, migration has been triggered by the rise of nationalism and emergence of new ethnically, religiously, and linguistically divided postcolonial states. Many governments attempted to decolonize their states by pressing for the migration of people who had been imposed on them by imperial powers. In attempts to expel their economically successful minorities, the government of Myanmar adopted discriminatory policies toward its Indian population while Indonesia and Vietnam did the same toward their Chinese populations. The Khmer Rouge viewed educated citizens as class enemies hostile to the state, and its extermination policies produced a mass migration of refugees. Similarly, the Vietnam War and subsequent invasion of Cambodia triggered a wave of refugees that contributed to significant regional conflict.

Today, Southeast Asia is largely at peace and, with the important exception of the Rohingya refugees who are discussed below, migration is overwhelmingly driven by economic factors. Neoclassical economic theory conceptualizes migration as a means of redistributing labor.[2] Population growth, poverty, and uneven economic development are regarded as key factors driving economic migrants, as are wage differentials, employment opportunities, and other market forces. Workers in countries such as Indonesia, the Philippines, Myanmar, and Cambodia migrate to seek employment in labor-scarce neighbors such as Malaysia and Singapore. Migration occurs through formal channels regulated by the state as well as through informal illegal movements of workers that occur outside of state channels.

Labor migration, particularly when regulated through formal channels, should be an area of cooperation between states. Countries where labor is scarce such as Malaysia and Singapore can benefit economically by employing economic migrants from labor-rich countries such as Indonesia and the Philippines. Efforts to capture the mutual benefits of economic remittances and employment opportunities for labor-sending countries and the labor productivity for receiving countries that are at the heart of this exchange form the basis for many bilateral Memorandums of Understanding (MOUs) to regulate migration.

Labor migration, however, is also a source of conflict between sending and receiving countries over issues such as changing economic conditions, working conditions, salaries, abuse, and criminal activities. Labor migrants who are welcomed or at least tolerated by receiving countries when economies are booming may be viewed as threats to local nationals when economic conditions deteriorate. Efforts by governments to protect their migrant workers from exploitative labor conditions in neighboring countries conflict with demands from employers in receiving countries to privilege the interests of their own citizens and firms.

In this chapter, we explore the interests and concerns of labor-sending and labor-receiving countries primarily through an analysis of migrant workers in the Indonesian-Malaysian relationship. We chose to study this relationship because of its scope, diversity, and changes over time, but many of the problems over migrant workers occur regardless of country of origin. We find that labor migration triggers conflict when it shifts from being perceived as an economic issue that produces gains for both countries to one that becomes defined as a political and security issue in which the interests of sending and receiving states are viewed as threats to one another. Indeed, the issue of migrant workers has become so politically contentious that Indonesia and Malaysia have each taken steps to stop it: Indonesia by issuing a ban on female citizens migrating to work as domestic helpers, and Malaysia by issuing decrees to ban its employers from hiring migrant workers. We find that better domestic governance of migrant workers in both countries could help ameliorate conflict in the relationship, but neither side has been willing to take on powerful vested interests to do so.

In this chapter, we explain that international and regional efforts to promote cooperation on migrant workers are weak. Efforts to protect migrant workers by outlining a set of migrant worker rights are codified in the International Convention on the Protection of the Rights of All Migrant Workers and Members of Their Families.[3] The convention lays out a standard for the treatment of migrant workers to guide policy, including rights to freedom of religion, a fair wage, and freedom from servitude and arbitrary determination.[4] To date, however, it has been signed by fewer than fifty countries, with no major Western or Southeast Asian labor-receiving countries among its signatories. Instead, it has been signed largely by labor-sending countries, with the Philippines and Indonesia being the only Southeast Asian signatories. Given the conflicting interests between source and destination countries, and the localized impacts of migrant workers, global governance is weak on this issue. Similarly, efforts to strengthen cooperation at the regional level have resulted in a number of ASEAN declarations on this issue, but no binding agreements. The same divergence of interests that stymie cooperation at the international level hinder it at the regional level as well.

In this chapter, we also examine the Myanmar government's persecution of the Rohingyas, which has produced an exodus of people and triggered conflict in Southeast Asia. According to the UN High Commissioner for Refugees, in 2015—a year that witnessed a record number of stateless people, refugees, and displaced persons—Myanmar ranked as the eighth-largest source country of refugees, with 451,800.[5] The Rohingyas meet the standards to be classified as refugees and, as discussed below, should therefore be granted temporary asylum and protection by receiving states. Instead, many Southeast Asian states have adopted beggar-thy-neighbor policies of pushing the Rohingyas back to sea and attempting to avoid paying the social and economic costs of hosting them. The Rohingya issue, therefore, is triggering conflict not only between Myanmar and its neighbors, but also among the Southeast Asian states to which the Rohingyas are fleeing.

In this chapter, we examine global and regional efforts to promote cooperation toward the Rohingyas. Internationally, refugees are the types of migrants with the greatest legal protection as outlined in the 1951 Convention Related to the Status of Refugees and its 1967 Protocol. According to the convention, a "refugee" is someone who

> owing to well-founded fear of being persecuted for reasons of race, religion, nationality, membership of a particular social group or political opinion is outside the country of his nationality and is unable or, owing to such fear, is unwilling to avail himself of the protection of that country; or who, not having a nationality and being outside the country of his former habitual residence as a result of such events, is unable or, owing to such fear, is unwilling to return to it.[6]

In addition to the rights of refugees, the obligation of receiving states to offer sanctuary to refugees while they apply for asylum, await resettlement, or wait for a change in conditions that permit a voluntary return to their country of origin are also codified in global governance relating to refugees. Southeast Asian states, however, have traditionally been reluctant to ratify the Convention Related to the Status of Refugees and provide even temporary shelter to refugees. Despite this reluctance, Southeast Asian countries and the international community were ultimately able to mount a cooperative response to the Indo-Chinese refugee crises in the 1970s and 1980s. In this chapter, we briefly compare the international and regional response to the Rohingya refugees with that of the Indo-Chinese refugees. We find that a willingness on the part of wealthy Western states to accept Indo-Chinese refugees for resettlement was a key factor in promoting cooperation toward those refugees. Today, compassion fatigue among the wealthy states that normally accept large numbers of refugees means that the vertical coordination between Southeast Asian states and other global actors in the fragmented refugee governance architecture has broken down. In the absence of leadership by outside states willing to offer permanent resettlement to the Rohingyas, or the abandonment by the Myanmar government of its persecution of the Rohingyas, the issue will likely continue to be a source of conflict between Southeast Asian states.

Indonesian Migrant Worker Issues: Background and Economic Significance

The practice of sending Indonesian workers abroad dates back centuries to the Dutch colonialists, who sent Indonesians to work in other colonies such as Suriname or Caledonia or to the Netherlands itself under a system known as "kuli kontrak" or contract labor.[7] The Indonesian government officially adopted its own policy through Decree No. 4/1970. Indonesia has always been plagued by poverty and overpopulation, particularly on Java. Sending Indonesians to work abroad was viewed as a means to absorb surplus labor, alleviate poverty, and generate foreign exchange, goals fully consistent with Suharto's overarching developmental strategy. In part because the Ministry of Manpower and Transmigration (MOM) was charged with overseeing migrant workers, these Indonesians were typically referred to as "migrant worker exports" illustrating the economic ethos of Suharto's New Order regime that ruled Indonesia for three decades until 1998 and sought to legitimize itself through economic development.

During the Suharto era, the government focused on maximizing the number of workers sent abroad to reduce unemployment and enhance remittances. Today, despite strong economic growth in recent years, almost 40 percent of Indonesians live on $2 a day or less, producing strong "push" factors for out-migration.[8] In 2014, the World Bank estimated that Indonesian migrant workers remitted a record $8.55 billion back to their families, an important source of income, particularly for the rural (mostly impoverished) areas from which they hailed.[9] According to the World Bank, remittances were equivalent to 0.9 percent of Indonesia's total GDP. It is estimated that each migrant worker supports five people in Indonesia. Obtaining precise data on the number of Indonesian overseas workers is problematic due to the myriad of organizations that play a role in the sending of workers abroad, each of which record statistics at different points in the process.[10] According to recent data from the National Agency for the Placement and Protection of Indonesian Migrant Workers (BNP2TKI), there are 4.3 million documented Indonesians working overseas, with approximately 1.2 million in Saudi Arabia and another 1.2 million in Malaysia.[11] Indonesia sends approximately 500,000 workers abroad a year through official channels.[12] Women account for almost 80 percent of Indonesia's official migrant outflows, with most working in the domestic sector as maids and caregivers. Approximately one-third work in the Middle East, with Saudi Arabia ranking as the largest receiving country, while other top destination countries include Malaysia, Taiwan, Hong Kong, and Singapore.

Government statistics cover only documented workers, not those who are working overseas illegally. Estimating the number of illegal migrants is difficult because they are not accounted for in government statistics. The IOM estimates that the number of Indonesians working overseas is at least double the documented figure, which would make it 9 million people.[13] Other sources place the total number of Indonesians working abroad at an estimated at 6.5 million to 8 million, which means that an estimated 2–3 percent of Indonesia's population works abroad.[14]

As Michele Ford observes, the government has been criticized for decades for focusing much more on the remittances of migrant workers than on their safety and well-being.[15] Regrettably, many migrant workers find themselves victims of unscrupulous recruitment agencies, corrupt government officials, and abusive employers. Migrants face difficulties while working abroad due to the general disregard for labor laws by employers, if such laws even exist in the host country. The problems experienced by Indonesian workers overseas fall into two broad categories: those related to labor rights, and those related to violence.[16] Often, Indonesians go abroad after signing contracts agreeing to a specific job at a specified wage, only to find that the terms of the contract are violated in the destination country. Many Indonesians work in jobs deemed "dirty, dangerous, and demeaning" in which accidents are frequent. Employment agencies or employers may confiscate their passports, refuse to pay wages, refuse to grant days off, and force them to work excessive hours in unsafe conditions. Finally, migrant workers are often victims of violence, which can sometimes be the product of disputes regarding labor conditions.

Domestic workers are the most vulnerable to exploitation. There is little agreement about what constitutes a fair workload for a domestic worker, and law enforcement agencies are reluctant to monitor conditions in private homes. Domestic workers may be forced to work excessive hours and be denied food, opportunities to practice their religion, and contact with their families. Domestic workers may lack privacy, particularly if they sleep in public spaces such as the kitchen or hallway, as is common in places such as Singapore and Hong Kong. Domestic workers are dependent on the goodwill of their employers for their safety and well-being. But female migrants are often mentally, physically, and sexually abused. Some religious groups have criticized the morality of sending female workers overseas. As long as the Indonesian government viewed migrant workers in terms of the way in which they could promote the national economic interest, however, it downplayed these concerns. Defining Indonesia's interest in migrant workers in economic terms, rather than focusing on the well-being and human rights of individual workers, created the conditions for cooperation in Indonesia's relationship with Malaysia.

Indonesian Migrant Workers in Malaysia: From Welcomed Kin and Cooperation to Security Threat and Conflict

Migration between the two countries has gone on for centuries, but in recent decades has become what one authority calls "one of the most enduring problems for contemporary Malaysia-Indonesia relations."[17]

Historically, both British colonialists and Malay leaders welcomed Indonesian migration to meet the country's manpower needs. Due to shared racial, linguistic, and cultural similarities, Indonesian workers were favored by Malay rulers as demographic buffers against Chinese and Indian immigrants because the assimilation of the Indonesian migrants helped the Malays maintain their status as the majority ethnic group, which at the time of independence in 1957 was only slightly more than 50 percent. This political calculation grew stronger in the wake of the May 1969 race riots that sparked a reassessment of Malaysia's ethnic, political, and economic divisions and ultimately led to the adoption of the New Economic Policy, an affirmative action policy for the Malays. As the Malaysian economy began to industrialize and many Malays moved from rural to urban areas, Indonesian migrants took the agricultural jobs that the Malays left behind during the first major phase of migration into Malaysia since the two countries gained independence.

Of the 4.3 million documented Indonesians working overseas, some 1.2 million are in Malaysia, making it the second-largest destination country for workers arriving through documented channels.[18] In Malaysia, male migrant workers outnumber female ones. According to the IOM, the sectoral breakdown of Indonesian workers in Malaysia is as follows: 25 percent planation workers, 24 percent domestic, 18 percent construction, 17 percent manufacturing, 8 percent services, and 7 percent agriculture.[19] With regard to gender, women make up approximately 30 percent of migrants and work primarily in the domestic and services sector. Since many of the migrants work in the formal sector, they are covered by Malaysian domestic labor laws. In theory that should enhance their labor rights, thereby reducing the potential for exploitation and hence problems related to migrant workers in the bilateral relationship.

Given the geographic proximity, as well as the common language and culture, much of the migration from Indonesia to Malaysia is illegal. The ease of movement across borders on Kalimantan and the existence of kinship arrangements mean that strong informal and underground labor networks exist. Given the high recruitment fees paid by Indonesians seeking to migrate legally as well as the fees paid by Malaysian employers to Malaysian labor recruiters, both workers and employers have a financial incentive to use illegal channels rather than formal ones. This raises the potential for migrant workers to become problems in the bilateral relationship.

The second phase of massive migration began after 1991, when the Malaysian government announced that foreign workers could be recruited in the plantation, construction, and domestic service sectors. The number of documented overseas workers, estimated at 1.0 million in 1990, rose to 1.5 million in 1995 and close to 2.0 million when the Asian financial crisis struck in 1997, in addition to an estimated 1.0 million undocumented workers. The third phase of labor migration began following the recovery from the Asian financial crisis and reached a high of 3.0 million documented migrants by the end of 2008, with an estimated 800,000 undocumented workers. The number of migrants fell to 1.9 million following the 2008 economic downturn, according to the Malaysian government.[20] Indonesians today make up more than half of the foreign labor force, with an estimated 1.2 million documented workers and an estimated 1.0 million undocumented workers.[21]

Over time, as Joseph Liow argues, Malaysian perceptions of Indonesian migrants changed dramatically. When an increase in Malaysia's crime rate coincided with a growing influx of Indonesian workers, Malay leaders began to view Indonesian migrants not as ethnic kin to be welcomed but as criminals to be feared.[22] As the number of illegals grew, do did competition for jobs, particularly those traditionally dominated by ethnic Malays.[23] Furthermore, as it became known that a large number of Indonesian migrants were Christians who had begun proselytizing among the Malay community, one Malaysian cabinet minister called the potential conversion to Christianity "the biggest threat facing Muslims in Malaysia today."[24] In short, by 2001 a combination of crime and economic and religious conflicts led to a new political discourse in Malaysia that viewed Indonesians as threats to national security.

In response to the growing problems among its large migrant labor population, the Malaysian parliament passed legislation in October 2001 that limited work permits to three years, in contrast to the previous six-year term. As a result, many Indonesian workers who had been in the country for longer than three years on official six-year work permits found their official status changed from legal to illegal overnight. Furthermore, the government also announced its intention to repatriate 10,000 illegal workers a month, and began detaining newly illegal workers in detention centers that had often been criticized as overcrowded and unsanitary. When detained Indonesians sparked a series of riots in which some Malaysian police suffered injuries, public opinion in Malaysia hardened further against the Indonesians. Under pressure, the Malaysian government responded by issuing new corporal punishment regulations that instituted flogging for first-time immigration offenders and the immediate repatriation of detained illegal workers. In January 2002, two separate violent riots by Indonesian workers, one involving 400 workers at a textile factory who torched buildings, reinforced Malaysia's perception of Indonesians as a threat to the country's security.

Until this point, Indonesia and Malaysia had managed to keep the migrant worker issue from negatively impacting other aspects of their important bilateral economic and security relationship. This was in part a function of the Indonesian government's framing of the migrant worker issue as an economic one. Jakarta's interest in ensuring the continued migration of Indonesian workers to Malaysia had traditionally led government officials to refrain from criticizing Malaysian actions. Although Malaysia had ordered a series of mass deportations of undocumented workers in the 1990s, one authority argues that there was little evidence of Indonesian protests in response.[25] When Malaysia deported 160,000 undocumented workers in 1997–1998, mostly Indonesian and Filipino, the Indonesian ambassador announced that Jakarta was satisfied with how the Indonesian detainees were treated.[26] Even in 2002, the Indonesian ambassador to Malaysia apologized to the Malaysians for the January riots mentioned above.

By 2002, however, domestic politics on the migrant worker issue were changing in both countries. Political leaders in Malaysia and Indonesia were increasingly under domestic pressure to reframe the issue away from an economic one that was a basis for bilateral cooperation to a political, human rights, and security one that triggered conflict. On the Malaysian side, the minister of law, Rais Yatim, expressed the situation in these words: "Besides defying authority, they [Indonesian migrant workers] had the cheek to wave the Indonesian flag. They are not in Jakarta. They are in Malaysia. . . . Indonesia's Ambassador here need not say sorry anymore. We are going to take stern action. Malaysians in general cannot tolerate the violent behavior of the Indonesians who are being too extreme and ungrateful."[27]

In January 2002, the Malaysian government announced that it would embark on a policy of mass repatriation of illegal foreign workers. The same year, the Malaysian Immigration Act No. 1959/63 was altered so that working illegally in Malaysia was a criminal act, and foreigners doing so faced fines of $3,200 (10,000 ringgit), up to five years in jail, and six strokes of the cane or whip.[28] The repatriation policy applied to migrant workers of all nationalities, but Deputy Prime Minister Abdullah Badawi subsequently announced a "temporary halt" to the employment of Indonesian workers, and Prime Minister Mahathir Mohamad followed up by stating that it was time for Indonesian workers to be "replaced" by other foreigners.[29] Mahathir subsequently announced a Hire Indonesians Last policy in which Indonesians would be confined to work in the domestic sector and agriculture industries. Attempts to implement this policy, however, revealed Malaysia's dependence on foreign (particularly, Indonesian) illegal workers. It became clear that approximately 70 percent of construction workers were Indonesian, and 80 percent of them were illegal. As the Hire Indonesians Last policy was implemented, the number of construction workers dropped 40 percent, halting work on many projects and leading Malaysian businesspeople to call to an end to the policy. It was subsequently dropped.[30]

Malaysia may have dropped its Hire Indonesians Last policy but its clear anti-Indonesia intent, combined with disparaging treatment of Indonesian workers such as caning, triggered anti-Malaysian sentiment in Indonesia. In Indonesia's new democratic system, this in turn led to protests outside the Malaysian embassy in Jakarta and calls for a three-day labor strike by Indonesian workers. Illustrating a rise of Indonesian nationalism on this point, one newspaper article argued that "there was a time, not so long ago, when Indonesia did not take such a belligerent act from a neighboring country lying down."[31]

It was in this atmosphere of rising tensions in 2002 that Malaysia deported almost 140,000 undocumented Indonesian workers from the Malaysian state of Sabah across the border into the Indonesian town of Nunukan, where an estimated seventy workers died. Thousands more suffered serious illnesses such as malaria and dysentery because the Indonesian government was utterly unprepared to meet the needs of the returning workers, although it had foreknowledge of their impending arrival. Indonesians not only criticized the Malaysian government for repatriating the migrant workers, but also its own government for failing to provide sufficient resources to process the migrant workers and transport them to their hometowns. This 2002 humanitarian disaster generated an Indonesian backlash against Malaysia, but it also marked a turning point in Indonesian policy toward migrant workers because the country began to redefine its national interests in favor of the rights of migrant workers (as proponents of human security contend should be the case).

Redefining Indonesian National Interests on Migrant Workers: From Economics to Human Rights

The humanitarian disaster at Nunukan was also a watershed in Indonesian policy toward migrant workers because it created the political opening for advocates of migrant workers to galvanize public support for their efforts to strengthen government protection for migrant workers. Malaysia's 2002 decision to repatriate illegal Indonesian workers took place in the early years of Indonesia's reform or democratic era, which created new political freedoms for civil society groups to create NGOs, publicize the plight of migrant workers in Indonesia's newly free media, and lobby the parliament and other government officials to act on their behalf. The democratic opening created the political space for advocates of migrant workers to demand that the government incorporate the human security interests of migrant workers into national policy, rather than simply define the issue as an economic one.

Migrant advocates organized a class action suit against the government for its abject failure to provide essential public services to its citizens. The court recognized that the government was indeed negligent in its initial ruling, only to have it overturned on appeal. Despite its failure, this initiative provided an important precedent of migrant NGOs using the legal system to hold the government accountable to its citizens, something impossible in the authoritarian Suharto era.

Shortly after the humanitarian crisis at Nunukan, the migrant worker NGO community came together to lobby for a migrant worker law. Up to this point, the regulation of migrant workers was achieved by ministerial decrees and other government regulations that were much weaker policy instruments than legislation. Moreover, these decrees often contradicted one another and failed to deal comprehensively with the issue.[32] Within a year of the Nunukan crisis, three draft bills were being discussed in parliament, one of which was drafted by the migrant labor NGO consortium, Consortium of Indonesian Migrant Workers Advocacy (KOPBUMI). The bill passed in October 2004, marking a significant, but qualified, victory for migrant worker advocates.

In their campaigns to enhance protection of migrant workers, advocates often find themselves facing opposition from powerful well-funded actors who profit from migrant workers. Law No. 39/2004 included a provision in Article 11 stating that migrant workers could be sent only to countries with which the government of Indonesia had written agreements.[33] This provision was designed to enhance protection for the workers by enshrining in bilateral agreements a set of conditions to which both countries would abide. At the time, this provision was aspirational since the government had agreements only with Malaysia, South Korea, and Jordon.[34] Fearing a halt to their lucrative business, labor-sending companies launched a court appeal to have it amended.[35] To understand the domestic roots of some of the problems suffered by migrant workers abroad, as well as the powerful vested interests the NGOs faced in their advocacy campaigns, we discuss the domestic recruitment process next.

Law No. 39/2004 provides the legal framework governing the labor migration system. It focuses on governance and administration, the bureaucratic powers and responsibilities of different actors, licensing requirements for recruitment agencies, administrative requirements to travel abroad, and predeparture processes. Like many sending countries, Indonesia relies on private firms to recruit, train, and place migrant workers, which means that it has a hybrid form of migration governance. MOM has the principle responsibility for managing overseas labor migrants. It is tasked with setting standards to implement the law (including licensing and oversight of the private recruitment agencies), approving specific job orders from abroad, setting standards for all predeparture processes and documents, and supervising the placement of migrant workers.

In 2006, the president created the BNP2TKI to coordinate all of the relevant government actors involved in the migrant worker process to ensure that the placement and protection of workers is done in an integrated manner.[36] It reports directly to the president, not MOM. In contrast to MOM, BNP2TKI lacks any enforcement authority over the recruitment agencies or other actors involved in Indonesia's fragmented migration process. MOM, BNP2TKI, and the Ministry of Foreign Affairs (Kemlu) are the three key Indonesian government actors directly responsible for migrant workers.

The process of sending Indonesians overseas begins when Kemlu receives a job order from a placement agency in a foreign country detailing the number of workers needed, the job types, necessary qualifications, terms, and other particulars. Kemlu is responsible for reviewing the job order, and ensuring that the placement agency is reputable and has not been placed on a blacklist due to prior labor disputes. Once Kemlu conducts its due diligence, the job order is sent to MOM in Jakarta. MOM is responsible for approving a specific job, after which it issues a permit to recruit one of Indonesia's more than 500 private licensed recruitment agencies to fill the job order. These agencies handle the selection, training, departure, placement, and return of workers from overseas, giving them enormous influence in the migration process.

Many recruitment agencies, however, are located in the cities whereas potential migrants live in rural villages. Recruitment agencies therefore rely on local brokers to find and screen potential recruits. The recruitment agency is then responsible for conducting a competency test and physical and mental health exams. Once a recruitment agency has matched an applicant with a job order, a contract of work is signed, and the recruitment agency works with the migrant to complete the paperwork for a passport, a temporary work visa, and an insurance card. The recruitment agency is also responsible for providing a migrant worker with predeparture training if necessary and travel to the country of work.[37] Indonesia's recruitment governance is fragmented, involving a multiplicity of government, civil society, and for-profit actors active at the local and national level.

Indonesia's migration system is rife with potential for the exploitation of workers. The standards for licensing recruitment agencies are not onerous, and the process by which they are approved is opaque.[38] The

MOM is charged with overseeing the recruitment agencies, but in practice many bureaucrats are easily bribed to overlook a lack of compliance. Brokers are unlicensed and often exploit uneducated migrants. There are no standardized fees for brokers or recruiters, who have an interest in charging high fees. According to the IOM, the fee charged to work in Hong Kong is $1,600 (15 million Indonesian rupiah), an extraordinary amount given that the average worker's monthly salary in Hong Kong is $460 (3,500 Hong Kong dollars).[39] Migrants often borrow at usurious rates to pay the fee, giving them few options but to continue working to recoup the fee, even if they find themselves in difficult working conditions. Village heads interested in increasing local remittances lie about a migrant's age if they are younger than the required twenty-one years old. Recruiters promise well-paying regulated jobs that do not materialize and recruitment agencies fail to provide the stipulated training, thereby sending workers overseas to jobs for which they lack the requisite skills. As a result, many in Indonesia and abroad contend that the problems involved with protecting migrant workers overseas begin at home.[40] In short, Indonesia's recruitment process pits the interests of migrants against the recruitment agencies and corrupt officials who benefit from their exploitation. As a result, Indonesian migrants are often sent abroad without proper training, skills, and preparation, which helps lay the foundation for conflict with the countries where they are sent to work.

Redefining National Interests in Malaysia and Indonesia: Conflict over Migrant Workers

By the early 2000s, therefore, the way in which the migrant worker issue was being reframed in each country was also a source of conflict that made managing the migrant worker issue more difficult. In Malaysia, perceptions of Indonesian migrant workers were shifting from one of kin and economic benefit to security threats. This coincided with the Indonesian NGO community and like-minded politicians reframing the migrant worker issue from one of solely economics to one of rights as well. When Malaysia and Indonesia both viewed the migrant worker issue as an economic one that produced mutual, albeit unequal, benefits, there was significant scope for bilateral cooperation. But when Malaysia securitized the issue at the same time Indonesia was increasingly focused on migrants' human rights, the stage was set for greater bilateral conflict. As security fears led Malaysia to adopt policies permitting the use of corporal punishment, and authorizing civilian vigilantes to apprehend individuals suspected of being undocumented workers, there was an increase in the number of Indonesian citizens whose rights were violated. As Indonesian migrant worker advocates publicized these abuses, they triggered a rise in public opposition to Malaysia and demands that it enhance its protection of Indonesian migrant workers.

In May 2004, the severe abuse of an Indonesian domestic worker, Nirmala Bonat, triggered societal pressure on the Indonesian government and ultimately led to the signing of a bilateral MOU that had already been under discussion. The MOU covers migrant workers in the construction, factory, and plantation sectors, but fails to cover domestic workers, with the Malaysian minister of human resources indicating that a separate agreement would be needed.[41] The MOU provides greater protection for Indonesian workers through the issuance of identity cards, specifies that round-trip airfare should be paid jointly by Indonesian labor recruiters and Malaysian employers, and states that migrant workers should earn a minimum salary of $10 per day. However, the agreement still permits employers to hold passports, prohibits labor organizing, and fails to cover issues related to conditions of work. Therefore, many migrant worker advocates have criticized the agreement as too weak.[42] Indonesian government officials, however, claim that Indonesia felt inhibited from negotiating too hard because it feared Malaysia would turn elsewhere for cheap labor at a time when the Indonesian economy was still recovering from the Asian financial crisis.[43]

Despite the MOU, tensions rose in the mid-2000s as Malaysian policy toward illegal migrant workers combined periodic amnesties with harsh crackdowns. In an effort to flush out illegal workers, Malaysia announced in 2004 that illegal workers who left the country voluntarily during a four-month amnesty period would not be subject to fines, jail, or whipping penalties, but that it would launch massive operations against those remaining in the country.

In March 2005, the Malaysian government began Operation Firm, a nationwide hunt for illegal migrants. The manhunt was conducted not only by government officials but also by members of Ikatan Relawan Rakayat Malaysia (RELA), a civilian volunteer force that had been granted the authority to check travel documents and was authorized to arrest those without proper papers. In Operation Firm, RELA volunteers were given a bounty of $20 (80 ringgit) for each illegal immigrant arrested, leading to allegations of overzealous RELA volunteers using force against Indonesians.[44] This, of course, triggered condemnation from NGO activists. Despite the Malaysian government's importation of workers from Burma, India, Sri Lanka, and Pakistan, labor shortages were so severe that in May 2005 it permitted formerly illegal workers who had left under the amnesty to return to Malaysia to seek work.[45] The use of

RELA volunteers to take action against the threat posed by migrant workers illustrates the private-public partnership or hybrid governance in Malaysian domestic politics.

Tensions rose rather than declined in subsequent years. In 2006, the Malaysian government appointed RELA the lead agency in the fight against illegal immigrants, despite earlier criticism about its use of excessive force. To insulate subsequent crackdowns on illegal workers from the Indonesian press, Malaysia announced a media blackout on official operations. Even when RELA volunteers aggressively and mistakenly detained a number of prominent Indonesians with official documents (e.g., the wife of an embassy official and a karate coach in the country for an official competition), Malaysia chose not to make any formal apologies to Indonesia, which raised the level of conflict.[46]

Indonesians are resentful that even though workers are punished by caning or whipping, there have been few penalties imposed on Malaysian employers. Neither has the Malaysian government cracked down on the Malaysian labor importers who work with corrupt Malaysian officials to bring illegal migrants to the country in the first place. Employers have an incentive to employ illegal workers because they are cheaper and easier to exploit due to their fear of deportation. Just as Indonesian officials often find it more convenient to direct their anger over the mistreatment of migrant workers at foreign employers and governments rather than crack down on corrupt Indonesian recruitment agencies, brokers, and bureaucrats, so too does Malaysia prefer to penalize foreign workers rather than the domestic employers that exploit them. Cooperation would be enhanced if each side worked to eliminate illegal migrants, but this would require governments in both countries to crack down on private companies with the resources to lobby and bribe government officials.

Against this backdrop, a series of high-profile horrific cases of domestic worker abuse in 2009 led to public protests in Indonesia where the Malaysian flag was burned and demands for retaliation were made. In contrast to 2004, the Indonesian government in June 2009 banned new recruitment of Indonesian domestic workers, and stipulated that the ban would not be lifted until workers were permitted to hold their passports, granted a minimum wage, and given a weekly day off.[47] The Malaysian government agreed to negotiate along these lines, although it continued to issue visas to Indonesians seeking to work as domestic migrants, something Indonesian NGOs criticized.[48] The 2009 ban was partially lifted when the two countries signed a Letter of Intent stipulating that migrant workers could retain their passports and were entitled to a day off a week, but negotiations deadlocked over Indonesian demands that the minimum monthly salary be raised from $150 (500 ringgit) to $240 (800 ringgit).

The moratorium was officially lifted in June 2011 when the two countries signed a formal MOU that many migrant advocates argue failed to strengthen the position of migrant workers because it set no limit on recruitment fees, continued to allow employers to deduct the fees from the wages until paid, and failed to set a minimum wage.[49]

Protecting Indonesian Workers by Banning Migration

Protecting migrant workers was a hot button issue in Indonesia's 2014 presidential campaign that pitted Joko Widodo (aka Jokowi), a former furniture manufacturer and mayor of Solo, against Prabowo Subianto, a former general. Jokowi's Indonesian Democratic Party of Struggle (PDI-P) has traditionally been the staunchest supporter of migrant workers in the Indonesian parliament (DPR). Prioritizing the issue was consistent with the interests of the PDI-P key constituency among the lower-middle class and it was included in Jokowi's election manifesto. On taking office, Jokowi's administration explicitly adopted a pro-people foreign policy and Foreign Minister Retno Marsudi stated that Indonesian foreign policy over the next five years would focus on three main priorities: maintaining Indonesian sovereignty, enhancing the protection of migrant workers, and intensifying economic diplomacy.[50] Ranking the protection of migrant workers among the country's top three foreign policy priorities is a dramatic change in Indonesia's definition of its national interest, illustrating how a human security issue can become a national security issue.

Jokowi's February 5–7, 2015, trip to Malaysia provided a test of the importance of migrant worker protection to his administration. Jokowi's choice of Malaysia as his first bilateral trip to an ASEAN country—the same choice made by Susilo Bambang Yudhoyono a decade earlier—is an indication of the importance of the bilateral relationship. Malaysia is Indonesia's key economic partner in ASEAN, and Jokowi has made economic diplomacy a major tenet of his attempt to spur development at home. The Malaysian side hoped that Jokowi's visit would be an opportunity to focus on strategic issues

Shortly before the visit, the decision by a Malaysian vacuum cleaner producer to advertise its product with the "Fire Your Indonesian Maid NOW!" slogan triggered an uproar in Indonesia. MOM minister Hanif Dhakiri called the ad "insulting to our workers," Kemlu issued a formal complaint, and the Indonesian ambassador to Malaysia called on Malaysian authorities to ban the advertisement.[51] Indonesian vice president Jusuf Kalla called the ad "a form of harassment" and threatened a moratorium of Indonesian workers.[52] Rizal Sukma, a close foreign policy adviser to Jokowi, demonstrated the more nationalist tone in Indonesian foreign policy under Jokowi when he stated, "If Malaysia wants Indonesia to respect it, so it must respect our migrant workers, our culture, and our sovereignty."[53]

In the end, the uproar over the advertisement and the migrant worker issue did not appear to spill over into the broader relationship during the visit, which both sides deemed cordial. Jokowi met with 300 Indonesians working in Malaysia, among them professionals, academics, and low-skilled migrant workers, and pledged his commitment to assist them when they faced problems in Malaysia.[54] The two countries agreed to continue efforts to ensure that Indonesian domestic helpers were recruited through only official channels, not unofficial ones.[55] Malaysia agreed to permit the opening of a school for migrant workers.[56] In sum, the visit appeared to make small steps of progress on the migrant worker issue while preventing it from triggering conflict in the overall relationship.

Shortly after his return to Indonesia, however, Jokowi publicly stated that he felt very "ashamed" and "humiliated" when discussing the migrant worker issue with Malaysian prime minister Najib Razak.[57] In mid-February, Jokowi said that Indonesia would ban the sending of female domestic workers abroad, "to protect the nation's dignity."[58] Jokowi's plan to ban female migration was slammed by migrant worker advocates. Anis Hidayah, executive director of Migrant Care, Indonesia's most prominent migrant worker NGO, argued that "all citizens have the right to seek decent work, and it is the responsibility of the state to protect them no matter where they work."[59] Rather than infringe on the rights of workers by circumscribing their freedom of movement, NGOs called for the government to reform the institutional bureaucracy, particularly corrupt elements in MOM and BNP2TKI that had long exploited migrant workers. Further illustrating the domestic and international links of the migrant worker issue, NGOs called on the government to enact an Indonesian domestic workers law, which had been languishing in the DPR for years. Migrant worker advocates accused Jokowi of hypocrisy by demanding that foreign governments grant domestic workers rights they did not have at home. Leaders of districts dependent on migrant worker remittances called on the government to reconsider the ban.[60]

As the theoretical discussion of threats and vulnerabilities to citizens' well-being in Chapter 1 illustrates, danger to migrant workers is a function of their home state's inability to create enough jobs to satisfy needs at home, and the threats of exploitation from abusive employers and recruiters overseas as well as their vulnerability to these threats. Since danger can exist only when both threats and vulnerability are present, the Indonesian government's efforts to ban migrant workers can be viewed as an attempt to eliminate the vulnerability of its citizens by keeping them away from threats. This will work, however, only if individuals banned from migrating can find jobs at home. Migrant worker advocates contend this is not the case. They argue that banning female Indonesian citizens from working overseas will have perverse consequences in the absence of a sufficient supply of jobs or other steps to alleviate poverty at home. Nusron Wahid, head of BNP2TKI, has stated that with the Indonesian economy growing at 5 percent, only 1.5 million of the 2.8 million new entrants to the job market can secure work at home, which will force others to do so abroad.[61] Eni Lesteri, chairwoman of the International Migrants Alliance, contends that the "moratorium makes women more vulnerable because they will migrate anyway."[62] Advocates argue that the ban puts Indonesian women at greater risk of human trafficking since they will now travel abroad illegally, with unscrupulous middlemen.

Lawmakers in the Indonesian parliament echoed these concerns. In a September 2015 meeting with the MOM minister Hanif, DPR members urged the government to lift the ban, claiming that it was never meant to be permanent and that it was forcing workers to travel illegally. DPR members argued that the ban was to be a temporary measure designed to provide the government with time to institute a new migration system that would enhance protection of migrants. As Ketut Sustiawan from the PDI-P argued, "Where's the next step in terms of improving protection? Don't let the ban encourage the illegal migration of Indonesian workers instead."[63] One move to enhance protection of migrant workers is a plan for the BNP2TKI to set up representative offices abroad in major destination countries, something that the Philippine Overseas Labor Office (POLO) initiated. The Philippine system, which many believe is one of the strongest and most effective, has emerged as a model that many advocates for migrants, the BNP2TKI, and others in Indonesia are studying as they seek better ways to manage the migrant worker system and better protect workers. Trade Minister Thomas Lembong has suggested that Indonesia should seek to include stringent contract terms for migrant workers in future trade deals since services are an increasingly important aspect of trade deals and migrant workers provide services.[64] These new mechanisms have the potential to enhance protection, but will do so only for workers traveling through official legal channels.

If banning workers is an attempt to reduce their vulnerability, then pressing the Malaysian government to create incentives for Malaysian employers to treat migrants more fairly is a way of reducing the threat to migrants. In early 2016, MOM Minister Hanif met in Kuala Lumpur with Deputy Prime Minister Ahmad Zahid Hamidi to advocate an increase in the minimum wage for domestic workers to $272 (1,200 ringgit) from the current average of $227 (1,000 ringgit) and for employers to bear the placement costs. According to Hanif, the Malaysian side hinted at support for the wage increase.

Shortly thereafter, however, Malaysia adopted a unilateral tax levy increase on overseas workers outside the domestic sector in an effort to incentivize its companies to move up the value chain and away from dependence on low-wage manufacturing, plantation, and service work. Effective February 1, 2016, each worker in the manufacturing and services sector would have to pay a levy of $591 (2,500 ringgit) versus the earlier levy of $296 (1,250 ringgit).[65] Levies for foreign workers in the plantation and agricultural sectors will rise from $140 (590 ringgit) to $355 (1,500 ringgit). When the levies on foreign workers were instituted in 2009, they were designed to be paid by employers, creating an incentive to hire domestic workers rather than foreign ones. In practice, employers forced the employees to bear these costs. Therefore, when Malaysia raised the levies that migrant workers must pay—and raised them on workers already in the country, not just new ones—it added a new tax on workers who had made their work plans based on a different set of fees and taxes. The rise in levies adds a new source of contestation between Indonesia and Malaysia.

Finally, in mid-February 2016, the Malaysian government suspended the recruitment of all foreign workers into the country while it reviewed the two-tier levy after it had received complaints from both Malaysian business owners and migrant worker advocates. According to Deputy Prime Minster Hamidi, "We urge all employers to recruit local workers" and advised young Malaysians to take advantage of the opportunities created by the ban.[66] Rather than attempt to redress the conditions of migrant workers, Malaysia is seeking to eliminate the problem altogether.

The Indonesian and Malaysian governments, therefore, are attempting to use the power of government to halt migration as it increasingly is perceived as a threat to the ruling elites' definition of national interests. In contrast to the argument that migration is a source of cooperation, it has become a source of conflict and risen to such an extent that both sides perceive national interests as better off if it is curtailed. The fragmented nature of the migration governance system in Malaysia and Indonesia, as well as the interests of unscrupulous employers, labor recruiters, and corrupt officials at all levels of government in each country, however, make it exceedingly unlikely that migration will be halted. Instead, the inability to forge a more cooperative formal migration governance structure will likely increase illegal migration and, ultimately, exacerbate conflict.

ASEAN Efforts to Promote Cooperation on Migrant Workers

ASEAN has made some modest efforts on the issue of migrant workers in Southeast Asia. It is an issue whose saliency is likely to grow. Some member countries will undergo a shrinking workforce (Thailand, Singapore, and Vietnam) while others will continue to grow (Indonesia and Cambodia); thus, the push and pull factors driving migration will continue.[67] ASEAN could play a significant role in bridging the differences between sending and receiving countries by promoting the adoption of regional standards for the recruitment, placement, and working conditions to enable them to capture the potential gains from labor migration. Heretofore, it has fallen short of this goal due to conflicting interests among regional countries and because effectively addressing the migrant worker issue requires changes in domestic governance in Southeast Asian states, something that butts up against ASEAN's norm of noninterference.

In 2007, ASEAN agreed to the Declaration on the Protection and Promotion of the Rights of Migrant Workers and established the ASEAN Committee to Implement the Declaration on the Protection and Promotion of the Rights of Migrant Workers (ACMW) to put it into practice. The goal is to increase codification, governance, and protection of rights for migrants working overseas, and to help fight human trafficking in Southeast Asia.[68] Working with other international organizations like the IOM, International Labour Organization (ILO), and UN Development Fund for Women (UNIFEM), ASEAN officials have convened a yearly forum on issues related to the declaration. These meetings have been the source of recommendations to member states on providing information on safe migration for decent work to different stakeholders. The 2015 ASEAN Forum on Migrant Labour (AFML) focused on best practices for labor inspections and how to comply with labor laws to protect the rights of migrants. These meetings facilitate the sharing of information across countries and provide venues for negotiation. Sending states would favor a legally binding agreement for the protection of their workers overseas. But receiving states, such as Singapore, Malaysia, and Thailand, have resisted greater regulation and codification of such rights; thus, the declaration remains only a set of recommendations rather than a legally binding instrument.[69]

In the run-up to the creation of the ASEAN Economic Community (AEC) and ASEAN Socio-Cultural Community (ASCC) in 2015, sending countries such as Indonesia and the Philippines lobbied for the adoption of a formal binding migrant worker treaty, but this was rejected. The Social-Cultural Community Blueprint does lay out protections and rights for migrant workers, but it lacks enforcement mechanisms. The AEC has provisions for the free flow of skilled workers in eight sectors (e.g., doctors and accountants), but there is no mention of semi-skilled or low-skilled workers who make up the lion's share of migrants within Southeast Asia.[70] It is estimated that low-skilled and unskilled workers account for over 80 percent of ASEAN's migrant worker population. In contrast, it is estimated that the skilled workers covered under this agreement account for only 1 percent to 1.5 percent of migrant laborers in the region, which shows the privileging of protection for skilled rather than unskilled workers.[71] ASEAN's unwillingness to codify legally binding mechanisms in the area of migrant workers is consistent with its policies in other issue areas. ASEAN has long been criticized for its propensity to issue statements and declarations that it cannot implement. ASEAN's consensus decisionmaking rule means that any one of its ten members can veto policies that could enhance regional cooperation. Many ASEAN countries strongly guard their sovereignty and are prone to reject regional policies that would require them to make domestic adjustments to comply with those policies. Since the migrant worker issue is one that would require receiving countries to protect migrants from exploitation from their own citizens, it violates ASEAN's cardinal rule of noninterference in the domestic affairs of member states.

The Rohingyas: Refugees and Beggar-Thy-Neighbor Policies

The long-standing persecution of the Rohingyas, according to a group of ASEAN parliamentarians, has "led to the largest outflow of asylum seekers by sea since the U.S. war in Vietnam."[72] This grave migrant crisis has triggered significant regional conflict between Myanmar and its neighbors as well as between countries seeking to deny Rohingyas access to their states. In this section, we first outline the history of state and societal persecution of the Rohingyas in Myanmar, discuss the migration waves of the Rohingyas, and describe the response to them by individual Southeast Asian states. We then outline the global refugee governance architecture, the extent to which ASEAN countries abide by these conventions, and the efforts to promote collaborative solutions to the crisis within ASEAN and between ASEAN and other global actors. We briefly compare the current efforts to craft solutions for the Rohingya crisis to the earlier, more effective, global response to the Vietnamese refugee crisis. This comparison illustrates why cooperation is more difficult in the absence of a country, such as the United States, that is willing to help provide the economic costs of housing the refugees and, more importantly, make an offer of permanent resettlement.

Rohingyas are a group of Muslims, estimated to number 1.3 million, most of whom live in Rakhine state in Myanmar along the Bangladeshi border. Muslims make up 10 percent of Myanmar's population and are systematically marginalized from social, economic, and political life, but the Rohingyas have been singled out for special persecution.[73] Rohingyas were denied recognition as one of Myanmar's 135 ethnic groups at independence, and in 1982 the military junta introduced a citizenship law whose strict provisions stripped the Rohingyas of access to full citizenship. Further, in May 2013, local authorities in Rakhine state reaffirmed a ruling from 2005 that forbid Rohingyas living in particular townships from having more than two children per couple.[74] Other statutes require the Rohingyas to seek official permission to marry. Those who violate the law can face imprisonment.[75] Until recently, the government issued Rohingyas "white cards" that conferred some rights to freedom of movement, property rights, and access to education and employment. In 2015, the white cards were cancelled, and so were the rights they conferred.[76] Today, the Rohingyas are stateless.[77]

Beyond denying citizenship and the rights that come with it, the government has long used state violence, sometimes in coordination with other actors, against the Rohingyas. In 1991, a series of attacks on Rohingyas by local communities and security forces triggered the exodus of tens of thousands of them to Bangladesh. By 1995, many of these refugees had been repatriated by UNHCR. However, by 1996, many were again seeking to escape persecution in Myanmar and fled again to Bangladesh and Thailand, both of which did their best to refuse and turn back refugees.[78]

In 2012, anti-Muslim violence erupted when three Muslim men were arrested on suspicion of raping and killing a Buddhist woman, sparking massive riots and deadly clashes between Rakhine Buddhists and Rohingyas. Most of the victims of the violence were Rohingyas: estimates are that 280 Rohingyas were killed in Rakhine state and another 120,000 displaced.[79] Whole villages were burned to the ground and families who fled have not been allowed to return to their former homes. Human rights organizations documented evidence of security forces opening fire on the Rohingyas during the waves of violence, although this has been denied by the government. There is no doubt that, at a minimum, police and military personnel did little to quell the violence.

In 2013, anti-Muslim riots again broke out in major cities, killing more than seventy-five people with the government doing little to protect victims or punish perpetrators. The violence has been stoked in large part by the 969 Movement, a group that promotes a virulent form of Buddhist nationalism that amplifies racial and religious tensions. The organization advocates for the privileging of Buddhists in Myanmar and it encourages boycotts and vigilante actions against Muslims. One of the leading voices for these ideas is the charismatic Buddhist monk, Ashin Wirathu. Using the new press and political freedoms in Myanmar's nascent democracy, Wirathu delivers speeches and produces TV and radio messages that are filled with religious nationalist messages and anti-Muslim hatred. The implication from the 969 Movement and others is that Buddhism is under siege in Myanmar and that the monks, the state, and ordinary Buddhists need to be vigilant against Muslims and other non-Buddhists who want to see other religious and cultural forms accepted as part of the national fabric.[80] Reflecting this ethnoreligious nationalism, the parliament recently passed a series of laws on religion and morality. Since 2012, the anti-Muslim violence, propaganda, and laws have caused 140,000 Rohingyas to become IDPs and another 120,000 to seek safety abroad.[81]

Some Rohingyas fled by land to Bangladesh, which hosts more than 32,000 registered refugees; another 200,000 unregistered Rohingyas are estimated to be in the country, many in squalid camps. Others have paid smugglers to transport them out of Myanmar by boat. An estimated 80,000 fled from the Bay of Bengal since January 2014.[82] By September 2014, more than 137,000 refugees from Myanmar were registered in Malaysia, including tens of thousands of Rohingyas.

Thailand is a hub for regional human smuggling activities. During 2012–2015, many refugees paid smugglers to get them to Thailand, where they would travel by foot to Malaysia or board other boats to Malaysia or Indonesia. In its 2014 "Trafficking in Persons Report," the US State Department downgraded Thailand to Tier 3, the bottom rank, as a source, destination, and transit country for trafficking.[83] This designation, and the discovery of mass graves in what was believed to be a smuggling detention camp, led the Thai government to crack down on smuggling. Fearful of being caught, smugglers started abandoning their ships at sea.

In the spring and summer of 2015, an estimated 5,000–8,000 people were left stranded at sea with Southeast Asian states refusing to let them dock. Tan Kok Kwee, first admiral of Malaysia's Maritime Enforcement Agency summed up the prevailing sentiment as, "We won't let any foreign boats come in."[84] Instead, the navy planned to provide them with provisions and send them on their way. Almost simultaneously, Indonesia pushed back a boat packed with Rohingyas after providing water, food, and directions to Malaysia. Clearly, if Myanmar's efforts to cleanse its country of Rohingyas can be viewed as creating a huge negative externality for its neighbors, then rejection of Rohingyas by Southeast Asian states indicates a strong desire to avoid paying the economic, social, and political costs of accepting them. The United Nations responded by pleading with Southeast Asian states to open their borders to prevent further suffering, and human rights activists slammed the "not in my backyard" policies.[85]

Global Governance on Refugees and Regional Policy Toward Rohingyas

At the global level, refugees are the type of migrants with the greatest legal protection as outlined in the 1951 Convention Related to the Status of Refugees and its 1967 Protocol. According to the convention, a "refugee" is someone who

> owing to well-founded fear of being persecuted for reasons of race, religion, nationality, membership of a particular social group or political opinion is outside the country of his nationality and is unable or, owing to such fear, is unwilling to avail himself of the protection of that country; or who, not having a nationality and being outside the country of his former habitual residence as a result of such events, is unable or, owing to such fear, is unwilling to return to it."[86]

By this definition, the Rohingyas are refugees and entitled to the key protection offered by the convention, which is non-refoulement or not being returned to the country of origin where they would face threats to their life. Once a person arrives in a signatory country and claims asylum or sanctuary, that state has an obligation to review their application. If a migrant is recognized as a refugee, then signatories have an obligation to provide for that person's well-being. The convention, therefore, can be viewed as one of the first international efforts to protect human security.

The Philippines is a signatory to the convention, but the Southeast Asian states most affected by the Rohingyas, such as Thailand, Malaysia, and Indonesia, are not. As a result, they have no formal obligation to accept asylum seekers. Given their proximity to Myanmar, however, they are termed "countries of first asylum," or the places where refugees seek protection temporarily and have their requests for asylum processed. UNHCR, the office charged with overseeing refugees, traditionally has proscribed three main policies toward refugees: naturalization and resettlement in host countries, voluntary repatriation, and resettlement in third countries. As discussed below, one of the key reasons that Southeast Asian states have been unwilling to provide sanctuary for the Rohingyas is that the prospects for voluntary repatriation or resettlement in third countries are so slim that they fear being left with the prospect of hosting them over the long term.

Southeast Asian responses to the Rohingyas were taken in part against the historical experience with the Indo-Chinese boat people. In the twenty years following North Vietnam's 1975 victory over South Vietnam and reunification under Hanoi's rule, UNHCR estimates that 840,000 people who fled the country by boat survived the trip and landed on foreign shores.[87] Approximately 200,000 arrived in Hong Kong, and most others in Southeast Asia. Then as now, countries like Malaysia pushed the boats back to sea, claiming they were Vietnam's secret weapons to destabilize the noncommunist countries of ASEAN. The international community responded by holding a series of high-level conferences to devise a collaborative division of labor between the countries of first asylum, the West, and UNHCR that would protect and process the refugees. The Southeast Asian countries of first asylum agreed to provide temporary sanctuary to the refugees while their cases were processed and they awaited resettlement in third countries. Seeking to protect their political and social stability from unwanted migrants, Indonesia and Malaysia opened refugee-processing centers on Galang Island and Sungei Besar respectively, isolated areas where the refugees could be easily controlled. UNHCR conducted the refugee screening, donor countries paid the expenses of hosting the refugees, and, most importantly, third countries agreed to accept them permanently. Eventually, Vietnam and the United States working with UNHCR created the Orderly Departure Program through which Vietnamese could apply directly to the United States for resettlement without leaving their country. Between 1975 and 1990, over a million and a half Indo-Chinese refugees were resettled in third countries, with the United States accepting the majority.[88] In the mid-1990s, the remaining 40,000 Vietnamese still living in Southeast Asian camps who had been denied refugee status and termed *economic migrants* instead were repatriated to Vietnam. In the Indo-Chinese case, the diverse set of actors in the fragmented refugee governance structure were able to cooperate to protect the refugees.

The Rohingya Refugee Crisis: Regional Conflict and a Breakdown in Global Refugee Governance

The 2015 wave of Rohingya refugees therefore paralleled the earlier exodus of Indo-Chinese boat people, not only by the fact that they washed up on Southeast Asian shores and were pushed back, but also because the international community pressed regional states to accept them. Like before, Indonesia and Malaysia were the countries most negatively affected. They argued that the root of the problem lay with the Myanmar government, that they lacked the resources to provide for the refugees, and that they had no obligation to accept them and strong national interests to reject them. In an effort to break the logjam and promote a cooperative resolution, the United States announced in May 2015 that it would be willing to accept Rohingya refugees and play "a leading role in any multilateral initiative" to end the crisis of the Rohingya boat people.[89] In response, Indonesia and Malaysia agreed to provide shelter to 7,000 refugees on a temporary basis while their applications for asylum were processed and they awaited resettlement. Thailand agreed to refrain from pushing boats back to sea.

On the surface, this cooperative effort to respond to the Rohingya boat people bears similarities to the earlier one that grappled effectively with the Indo-Chinese crisis, but a closer look reveals significant differences that are likely to stymie an effective concerted response. First, the agreement by Indonesia and Malaysia to shelter the refugees is contingent on eventual resettlement in third countries such as the United States. Moreover, it was made clear that assistance was short term, and would not be granted to others. Dewi Fortuna Anwar, a political adviser to the Indonesian vice president, alluded to the danger that accepting the refugees would only create incentives for others to follow: "We have to find ways of settling them as soon as possible without creating a new moral hazard."[90] Thus, this is a temporary fix to a specific group of refugees, not a comprehensive solution to the problem.

Second, this cooperation was predicated on a US commitment to accept Rohingyas. In the current political environment in the United States and elsewhere, where anti-immigrant and anti-Muslim sentiment is rife, it is doubtful that the United States will accept large numbers of Rohingyas. In contrast to the Vietnamese case in which the United States had a role in the conflict that produced the refugees and therefore a specific obligation to assist them, the United States has no such ties to the Rohingyas. In contrast, the United States has a greater responsibility for refugees fleeing Afghanistan and Iraq. In addition, the Rohingya crisis is occurring at a time when Europe, long a generous acceptor of refugees, is experiencing its own migrant crisis. Finally, the number of refugees resettled in third countries is extremely low: in 2012, a total of 88,578 refugees out of a total of 10.4 million recognized refugees worldwide were resettled in third countries.[91] For these reasons, it is difficult to see the global community accepting Rohingyas in large numbers. Voluntary repatriation is UNHCR's third method for dealing with refugees, and

40,000 Vietnamese were repatriated in the 1990s. As the following political discussion argues, however, it is difficult to see how this is possible at this time in the Rohingya case. To date, a key component of the international response to the Rohingya issue is to pressure the Myanmar government to treat the Rohingyas more humanely so that the incentive to migrate is reduced. For the ASEAN countries, the organization's numerous regional meetings provide opportunities to discuss these issues on the sidelines. At the same time, ASEAN's norm against noninterference constrains the group from a public robust critique of Myanmar's policies and demands that they be changed. Whether international pressure is effective in changing the government's policy toward the Rohingya will largely depend on how its political transition unfolds.

The increase in Rohingya migration occurred at the same time that Myanmar began to liberalize its politics. After almost fifty years of military rule, justified in part on the need to maintain stability in Myanmar's multiethnic state, national elections were held in 2010 in which the long-time opposition party, the National League for Democracy (NLD), was allowed to compete. The elections were far from free and fair, but in March 2011, the military handed power to a partially civilian government. The military still retains a great deal of power, but this marked the beginning of a significant series of political reforms. The press was freed, political activism from civil society actors including student groups was permitted, and political prisoners were freed, among other reforms. In 2012, Nobel Laureate, long-time political prisoner, and NLD chair Aung San Suu Kyi was allowed to compete in the 2012 by-election and win a seat in the lower house of parliament. General elections were held again in the fall of 2015 and the NLD won in a landslide. Aung San Suu Kyi is constitutionally barred from the presidency, but she serves as foreign minister and exerts strong influence on the government, something unthinkable a few years ago.

One would hope that an increase in political freedoms and democracy would result in increased tolerance and more humane treatment of minority groups like the Rohingyas, but that has not been the case. Instead, the upsurge in persecution of Myanmar's Muslim minorities is consistent with the "Voting Leads to Violence" thesis that the liberalization of political and civil freedoms, combined with the competition inherent in elections, creates incentives for majority groups to scapegoat minority ones.[92] Aung San Suu Kyi disappointed foreign governments and international human rights groups with her refusal to condemn the violence and discrimination against Muslims in general and the Rohingyas in particular. Aung San Suu Kyi and most party members feared that speaking out would hurt them at the polls, not help them. Monks are held in high esteem in Burmese society, even extreme ones preaching hatred. During military rule, the monks were one of the few voices of opposition against the military junta, and many monks were arrested and served time in jail for their political activities. This has given them a significant reservoir of support and goodwill in society, and any politician criticizing monks and the violence they advocated risks a major political blowback.[93]

In the spring of 2016, Aung San Suu Kyi began to speak out about the need to address persecution of the Rohingyas. Perhaps Myanmar may become more responsive to outside pressure to treat the Rohingyas more humanely now that the election period is over. A shift in policy toward the Rohingyas would reduce the push factor driving immigration and be welcomed by ASEAN and the international community. It is far from clear, however, whether Myanmar's political leaders like Aung San Suu Kyi are willing stand up to powerful domestic actors like the military and Buddhist monks to protect the security of a small minority group like the Rohingyas.

Conclusion

In this chapter, we illustrated that two distinct types of migration—labor migrants seeking economic gain and refugees fleeing political persecution—have triggered much greater conflict than cooperation in Southeast Asia. Both of these migration problems reflect failures of states to provide public goods, either in the form of an adequate job market or physical security and safety to the people living in their territory. The lack of regional cooperation to address the transnational nature of the problem also reflects weaknesses in regional and global agreements and institutions. Given that labor migration is a transnational issue involving the citizens of states that governments are created to protect, perhaps it should come as no surprise that migration triggers conflict rather than the cooperation that economists would suggest. Indeed, in contrast to issues such as global warming and pandemic disease that can approximate global public goods, economic migration is essentially a private good, fulfilling the needs of workers and employers rather than the good of society at large.

Examining the interests of domestic actors on each side yields some important insights in the factors that could reduce conflict and foster collaborative efforts. First, as the case of Indonesia illustrates, despite the government's increasing willingness to include the rights of migrant workers in its definition of national interests, it is unclear how far it will go toward equating the national interest with those of its citizens, and in whose interests it will act. If the government defined its interests fully in accordance with the social welfare of its citizens, as the human security literature argues it should, then the government would be expected to enact policies that maximize the protection of its citizens. Will it muster sufficient political will to pass a new labor law that replaces the current fragmented migrant worker recruitment system with one that transfers many of the responsibilities of the private actors in the current system to the government and NGOs, thereby reducing the profitability of migrant recruitment, enhancing training and capacity building at home, and reducing the chances for conflict abroad? Or, will it continue to privilege the economic interests of labor recruiters and corrupt officials?

Second, on the Malaysian side, the government similarly needs to decide in whose economic interests it will act. In contrast to official statements that curtailing migration will provide opportunity for Malaysian employment seekers, it is unclear if Malaysians will find those jobs desirable. Moreover, it is unclear whether the ban will last, as other attempts to curtail it have been overturned in the face of protests by employers in the low-tech sectors. If the ban is to stand, wages will need to rise to entice Malaysians to take those jobs, or the government will have to provide selective incentives for the transition to a new higher-skilled economy.

It should be noted that there is a tendency in the literature on cooperation on transnational security threats to assume that, once governments take steps to take citizens' welfare into account when crafting policy, cooperation is more likely to occur. This may not hold true in the case of migrant labor or refugees, as the Indonesian-Malaysian history of escalating tensions and the Rohingya case illustrate.

Notes

1. International Organization of Migration, "Key Migration Terms," 2011, www.iom.int/key-migration-terms.
2. Alan Dupont, *East Asia Imperiled: Transnational Challenges to Security* (New York: Cambridge University Press, 2001), 18.
3. UN General Assembly, *International Convention on the Protection of the Rights of All Migrant Workers and Members of Their Families,* Res. 45/158 (December 18, 1990).
4. Ibid.
5. United Nations High Commissioner for Refugees, "UNHCR Global Trends 2015," 2015, www.unhcr.org/en-us/statistics/unhcrstats/576408cd7/unhcr-global-trends-2015.html, 17.
6. United Nations High Commissioner for Refugees, *Convention and Protocol Relating to the Status of Refugees*, www.unhcr.org/en-us/3b66c2aa10.
7. "Overview of Indonesia's Overseas Worker's Protection," Indonesian Ministry of Foreign Affairs, www.kemlu.go.id/Documents/Overview%20of%20Indonesia.doc.

8. International Organization for Migration (IOM), *Labour Migration from Indonesia: An Overview of Indonesian Migration to Selected Destinations in Asia and the Middle East* (Jakarta: IOM, 2010), xii, www.iom.int/cms/en/sites/iom/home/where-we-work/asia-and-the-pacific/indonesia.html.

9. Anjaiah Veeramalla, "RI's Migrant Workers Remit Record $8.55 Billion in 2014: WB," *Jakarta Post*, April 20, 2015.

10. Palmira Permata Bachitar, "Chaotic Statistics of Indonesian Migrant Workers," *Jakarta Post,* January 26, 2012.

11. Nurfika Osman, "Jakarta, Riyadh Sign MoU on Migrants," *Jakarta Post,* February 19, 2014.

12. IOM, *Labour Migration from Indonesia.*

13. Prashanth Parameswaran, "The Trouble with Indonesia's Foreign Policy Priorities Under Jokowi," *The Diplomat,* January 9, 2015.

14. Bassina Farbenblum, Eleanor Taylor-Nickolson, Sarah Paoletti, *Migrant Workers' Access to Justice at Home: Indonesia* (New York: Open Society Institute, 2013).

15. Michele Ford, "Migrant Worker Organizing in Indonesia," *Asia and Pacific Migration Journal* 15, no. 3 (2006): 315–316.

16. IOM, *Labour Migration from Indonesia.*

17. Joseph Liow, "Malaysia's Illegal Indonesian Migrant Worker Problem: In Search of Solutions," *Contemporary Southeast Asia* 25, no. 1 (2003): 45.

18. Osman, "Jakarta, Riyadh Sign MoU on Migrants."

19. IOM, *Labor Migration from Indonesia,* 43.

20. Helen E. S. Nesadurai, "Malaysia's Conflict with the Philippines and Indonesia over Labor Migration: Economic Security, Interdependence and Conflict Trajectories," *Pacific Review,* 26, no. 1 (2013): 96.

21. IOM, *Labour Migration from Indonesia.*

22. Liow, "Malaysia's Illegal Indonesian Migrant Worker Problem," 49.

23. Malaysia sends its own workers abroad in more advanced industries than those staffed by Indonesians in Malaysia. In times of economic downturns, they are repatriated back to Malaysia, where they find Indonesians in their old jobs.

24. Liow, "Malaysia's Illegal Indonesian Migrant Worker Problem," 49.

25. Nesadurai, "Malaysia's Conflict with the Philippines and Indonesia over Labor Migration," 99.

26. Ibid., 100.

27. Liow, "Malaysia's Illegal Indonesian Migrant Worker Problem," 51.

28. M. Rezaul Islam, "Partnership Research Activities Between Malaysia and Indonesia: A Case of Labour Welfare," *Social Research Reports,* no. 25 (2013): 5; Nesadurai, "Malaysia's Conflict with the Philippines and Indonesia over Labor Migration," 100.

29. Liow, "Malaysia's Illegal Indonesian Migrant Worker Problem," 52.

30. Ibid.

31. Ibid., 53.

32. Ford, *Migrant Worker Organizing in Indonesia,* 318.

33. Ibid., p. 318, footnote 8.

34. Ibid. The lack of agreements was one of the reasons that labor recruitment and supply companies opposed the law.

35. Ibid., 318.

36. Wayne Palmer, *Indonesia's Overseas Labour Migration Programme* (Boston: Brill, 2016), chapter 2.

37. Farbenblum, Taylor-Nickolson, and Paoletti, *Migrant Workers' Access to Justice at Home,* 38–42.

38. Ibid., 42.
39. IOM, *Labour Migration from Indonesia,* xii.
40. Human Rights Watch, "Help Wanted: Abuses Against Female Migrant Domestic Workers in Indonesia and Malaysia," July 21, 2004, https://www.hrw.org/report/2004/07/21/help-wanted/abuses-against-female-migrant-domestic-workers-indonesia-and-malaysia.
41. Ibid.
42. Ibid.
43. Ibid.; Nesadurai, "Malaysia's Conflict with the Philippines and Indonesia over Labor Migration," 103.
44. Nesadurai, "Malaysia's Conflict with the Philippines and Indonesia over Labor Migration," 101.
45. Human Rights Watch, "Malaysia Country Summary," January 2006, www.hrw.org/legacy/wr2k6/pdf/malaysia.pdf.
46. Nesadurai, "Malaysia's Conflict with the Philippines and Indonesia over Labor Migration," 106; IOM, *Labour Migration from Indonesia,* 30.
47. Nesadurai, "Malaysia's Conflict with the Philippines and Indonesia over Labor Migration."
48. IOM, *Labour Migration from Indonesia,* 31.
49. Human Rights Watch, "Indonesia/Malaysia New Pact Shortchanges Domestic Workers," May 31, 2011, www.hrw.org/news/2011/05/31/indonesiamalaysia-new-pact-shortchanges-domestic-workers.
50. Parameswaran, "The Trouble with Indonesia's Foreign Policy Priorities Under Jokowi."
51. "Indonesia Files Complaint with Malaysia over Vacuum Cleaner Commercial," *Jakarta Globe,* February 4, 2015, The ad is visible in this source.
52. Prashanth Parameswaran, "Malaysia Comes Under Fire from Indonesia Ahead of Jokowi Visit," *The Diplomat,* February 15, 2015.
53. "Jokowi Set to Raise Thorny Issues in Malaysia Visit," *The Straits Times,* February 5, 2015.
54. "Jokowi Pledges to Protect Indonesian Workers in Malaysia," *Tempo,* February 7, 2015.
55. Shannon Teoh, "Malaysia, Indonesia Appoint Special Envoys to Handle Territorial Dispute," *The Straits Times,* February 6, 2015.
56. "Jokowi, Najib, to Appoint Envoys on Boundary Row," *Jakarta Post,* February 7, 2015.
57. Indra Budiari, "Domestic Workers' Conditions at Home No Less Worrying," *Jakarta Post,* February 16, 2015.
58. ASEAN Trade Union Council, "Indonesia to Impose Ban on Maids Working Overseas Earlier than Expected," February 23, 2015, http://aseantuc.org/2015/02/2122015-indonesia-to-impose-ban-on-maids-working-overseas-earlier-than-expected/.
59. "Jokowi's Ban on Migrant Workers Unconstitutional: NGO," *Jakarta Post,* February 15, 2015.
60. "Government Urged to Reconsider the Ban," *Jakarta Post,* February 15, 2015.
61. Khoirul Amin, "Indonesia Seeks Greater Int'l Access for Its Workers," *Jakarta Post,* January 13, 2016.
62. Astrid Zweynert, "Indonesia's Maid Ban Increases Risk of Human Trafficking: Campaigner," Reuters, June 17, 2015.
63. Edi Hardum, "Legislators Seek End to Ban on Migrant Workers Going to Middle East" *The Jakarta Globe*, September 3, 2015.
64. Amin, "Indonesia Seeks Greater Int'l Access for Its Workers."

65. Vanitha Negedraj, "Levy for Foreign Workers: Who Will Blink First?" *ASEAN Today,* www.aseantoday.com/2016/02/levy-for-foreign-workers-in-malaysia-who-will-blink-first/.

66. "Malaysia Halts Recruitment of All Foreign Workers," *Asia Times,* February 19, 2016.

67. ASEAN Trade Union Council, "Migrant Workers in ASEAN: The Hidden and Neglected Workforce" (ASEAN Trade Union Council, May 26, 2015).

68. Human Rights in ASEAN Online Platform, "ASEAN Committee on Migrant Workers," http://humanrightsinasean.info/asean-committee-migrant-workers/about.html.

69. Abdulkadir Jailani, "The Need for an ASEAN Treaty on Migrant Workers," *The Diplomat,* January 22, 2015.

70. Andy Hall, "Migrant Workers' Rights to Social Protection in ASEAN: Case Studies of Indonesia, Philippines, Singapore and Thailand" (Migrant Forum in Asia and Friedrich-Ebert-Stiftung, 2011).

71. Wasamon Audjutarint, "ASEAN Struggles to Protect Migrant Workers," *The Nation,* May 30, 2016.

72. "The Rohingya Migrant Crisis," Council on Foreign Relations Backgrounders, June 17, 2017, www.cfr.org/burmamyanmar/rohingya-migrant-crisis/p36651.

73. Joshua Kurlantzick and Alyssa Ayres, "Sectarian Violence in Myanmar," 2017, www.cfr.org/global/global-conflict-tracker/p32137#!/conflict/sectarian-violence-in-myanmar.

74. Paul Armstrong, "Q&A: What's Behind Sectarian Violence in Myanmar?" CNN, October 26, 2012.

75. International Crisis Group, "Myanmar: The Politics of Rakhine State," Asia Report No. 261 (International Crisis Group, October 22, 2014); International Crisis Group, "The Dark Side of Transition: Violence Against Muslims in Myanmar," Asia Report No. 251 (International Crisis Group, October 1, 2013); Freedom House, "Freedom in the World: Myanmar, 2014 Report," http://freedomhouse.org/.

76. "The Rohingya Migrant Crisis."

77. The government and most people in Myanmar refuse to use the name "Rohingyas." Instead, these people are referred to as "Bengalis" by the government, which implies of course that they come from Bangladesh; thus, the government feels justified in arguing that they are not citizens of Myanmar and should go elsewhere. The Rohingyas have resided in this part of Myanmar for generations and it is believed that most came from parts of India (and parts of what would become Bangladesh) when the region was under British colonial control. For more information see International Crisis Group, "Myanmar."

78. Human Rights Watch Asia, "Burma, the Rohingya Muslims: Ending a Cycle of Exodus?" (New York: Human Rights Watch Asia, September 1996).

79. "The Rohingya Migrant Crisis."

80. Alexandre Marchand, "Inside the 969 Movement," *Ruom,* June 8, 2013.

81. Council on Foreign Relations, "Global Conflict Tracker," accessed October 24, 2017, www.cfr.org/interactives/global-conflict-tracker#!/conflict/sectarian-violence-in-myanmar

82. "The Rohingya Migrant Crisis."

83. US Department of State, "Thailand: 2014 Trafficking in Persons Report," https://www.state.gov/j/tip/rls/tiprpt/countries/2014/226832.htm.

84. Eileen Ng, "Worries Grow for Rohingya, Bangladeshi Migrants as Malaysia, Indonesia, Refuse to Accept Them," *Canadian Press,* May 13, 2015.

85. Kanyakrit Vongkiatkajorn, "The Year Southeast Asia Faced Its Own Refugee Crisis," 2015 Year in Review, *Vice News,* December 27, 2015.

86. "Text of the 1951 Convention Relating to the Status of Refugees and Text of the 1967 Protocol Relating to the Status of Refugees," Article 1, "Definition of the Term 'Refugee,'" 14, http://www.unhcr.org/3b66c2aa10.pdf.

87. "UNHCR and Malaysia Close Camp for Vietnamese Boat-People," June 25, 2001, www.unhcr.org/en-us/news/press/2001/6/3ae6b81838/unhcr-malaysia-close-camp-vietnamese-boat-people.html.

88. D. R. SarDesai, *Vietnam: Past and Present* (Boulder, CO: Westview Press, 1998), 173.

89. Jamil Maidan Flores, "The Rohingya Need a Lifeline," *Jakarta Globe,* May 25, 2015.

90. "Indonesia and Malaysia Agree to offer 7,000 Migrants Temporary Shelter," *The Guardian,* May 20, 2015.

91. Antje Missbach, *Troubled Transit: Asylum Seekers Stuck in Indonesia* (Singapore: ISEAS, Yusof Ishak Institute, 2015), 128.

92. Jack Snyder, *From Voting to Violence: Democratization and Nationalist Conflict* (New York: Norton, 2000).

93. Lawi Weng, "Support Incumbents, Ma Ba Tha Leader Tells Monks," *The Irrawaddy,* June 23, 2015.

7

Looking Forward: Prospects for Cooperation and Conflict

It seems that every day brings news about the transnational threats that we discussed in this book. Rohingya refugees rescued at sea in the spring of 2015 are interned in camps in Indonesia and Malaysia awaiting settlement in third countries willing to take them.[1] Haze from forest fires in Indonesia are casting a cloud over the region, albeit a much smaller one than in 2015. Rice crops are small because of drought conditions in major exporters such as Thailand and Vietnam; that drought is pushing small farmers off their land and sending them to more urban areas. Avian influenza is seemingly under control, but the Zika virus recently captured headlines in the Americas and Asia, reminding us that the next disease outbreak can occur in any place at any time. In this book, we looked at two critical questions concerning these challenges: How well are the problems being addressed? And how can we explain regional cooperation or lack of cooperation on these nontraditional security threats? How successfully or unsuccessfully people, communities, states, and regions cope with these stresses is a result of a complex mix of factors related to how they define and perceive these threats. At the domestic level, we find some successes and some failures. Likewise, we find some regional efforts to solve common problems, but not as much as one would hope.

In Chapter 1, our introduction, we distinguished human security (the well-being of the people living within a country) from national security (the well-being of the state itself). We also detailed three broad categories of national interests: (1) physical or territorial security; (2) economic security and prosperity (the well-being of the country and its citizens); and (3) the protection of values, civil culture, and a state's own system of rule. Climate change, food and water scarcity, pandemic disease, and migration all pose significant threats to the security of many people in Southeast Asia, and we argued that states are most likely to adopt effective policies to meet these challenges when they are perceived as posing significant threats to national security. When states define national security or the public good in ways that accord with human security, then they are likely to prioritize the problem, allocate resources to solve it, and muster sufficient political will to implement effective policies.

In Southeast Asia, we see some efforts at the national level of government to address problems of food and water security, climate change, the spread of infectious diseases, and migration, but in no case has the government of any state gone so far as to equate national and human security and pursue comprehensive security. Instead, competing interests make it challenging for states to act in the larger public interest. Privileged actors or groups are often powerful enough to get their way on key policies, thus making it harder for states to provide public goods. Weak state capacity in being able to carry out preferred policies and poor coordination between the national government and local officials, combined with problems of outright corruption, make policy implementation on the issues we discussed here difficult.

As we detail in Chapter 1, it is also important to understand how states perceive threats. Transnational problems do not pose the same degree of challenges to all of the Southeast Asian states. The magnitude of the threat, the degree of harm, and the likelihood of that threat vary from state to state. The most severe threats may also be the least likely to occur, and they may be the hardest problems to mitigate. Thus, leaders find it easier to address threats that are likely to materialize in the near term and are logistically easier to deal with than those that may occur further in the future.

Climate change, water and food scarcity, migration, and pandemic disease all pose significant challenges to Southeast Asian countries, but many of their negative consequences will be in the indeterminate future, which can reduce the incentives of states to respond to them now. Pandemics from emerging infectious diseases have a low probability of occurring, but could pose extremely high levels of threat to a state's interests. The immediacy of the threats posed by EIDs and food insecurity to the economic and social interests of Southeast Asian states helps us understand domestic efforts to resolve them. Climate change is already causing harm and has a high degree of probability of inflicting more dangerous consequences in Southeast Asia, but it is a hard threat to combat or stop. The challenges posed by these transitional issues are fairly intractable since there are no easy solutions to global warming or threats posed from infectious diseases. All of these factors often make it difficult for governments to respond effectively to transboundary threats in the near term.

At the regional level, we find some efforts to cooperate on transnational security issues, but in no case have Southeast Asian states been able to overcome the collective action problem and provide regional public goods. Southeast Asian states have repeatedly stated their interest in combating transnational threats and, in some cases, have adopted agreements and protocols in an effort to convert words into action. One barrier to more effective cooperation is the divergent interests among Southeast Asian countries that lead them to adopt beggar-thy-neighbor policies. The immediacy of the threat posed by a scarcity of rice to a country's food security may prompt governments to act domestically, but the differences in the interests of food exporting and importing countries help explain the relative lack of cooperative efforts to address food security regionally. Similarly, the competing interests of labor-sending and labor-receiving countries helps explain the relative lack of cooperation on migrant workers.

A second factor that helps explain the relative lack of regional cooperation is the weakness of regional institutions. Institutions, as discussed in Chapter 1, can help promote cooperation by increasing information and transparency, lowering the transaction costs of negotiating agreements, and devising compliance mechanisms to hold states accountable to their commitments. Regional institutions such as ASEAN and the Mekong River Commission have played important roles in sharing information and increasing transparency, and in the case of ASEAN in reaching agreements, but both suffer from weak leadership and are hamstrung by strong norms of state sovereignty that make it difficult to compel states to uphold their commitments. ASEAN has helped disseminate information on climate change and its members have committed to the ASEAN Agreement on Transboundary Haze Pollution, but the organization has no formal means to force Indonesia to crack down on the activities that produce the haze, despite the tremendous health and economic costs it imposes on neighboring countries. Similarly, the Mekong River Commission has produced valuable scientific studies on the river and changes in its flow, but the organization has no mechanism to compel members such as Laos to factor the interests of downstream countries into its decisions on dam building. China's refusal to join the MRC, and its decision to create a new institution in which it will write the rules according to its own interests, shows that powerful countries with an interest sufficiently large to alter outcomes exist, but in this case the country is using this capacity for its narrow national interest instead of the broader regional one.

A final obstacle to effective regional cooperation on transnational security threats requires that states, institutions, NGOs, and private actors must come together to overcome the challenges of fragmented governance. The multiplicity of actors involved in solving these complex problems makes it more likely to see failure instead of success. When international or regional organizations such as the World Health Organization and ASEAN develop mechanisms for monitoring and responding to disease outbreaks, the implementation of these agreements requires horizontal coordination between these institutions, Southeast Asian states, and NGOs as well as vertical coordination between these actors and local-level officials. There are many places where compliance failures can occur, which shows the need to promote networked cooperation among the different types of actors at different levels of governance. The hybrid and fragmented nature of the governance of transnational security challenges demonstrates how complex these challenges are to address and why cooperation is hard to achieve.

Table 7.1 summarizes our findings about the efficacy of Southeast Asian efforts to respond to the transnational security challenges of climate change, food and water security, pandemic disease, and migration. We assessed domestic governance by the extent to which states have the capacity to create policies to address transnational security issues and compel compliance with these policies so that the broad national interest is served. We ranked domestic governance strong if at least some Southeast Asian states had the capacity to create and compel compliance with policies that effectively respond to transnational security threats, moderate if states had some capacity to devise and compel compliance with polices, and weak if states had little capacity to either devise policies that would effectively meet these challenges or enforce compliance with them.

Similarly, we assessed the efficacy of regional and global institutions according to their ability to create and enforce treaties, conven-

Table 7.1 Efficacy of Southeast Asian Responses to Transnational Security Threats

	Food Security	Climate Change	Health Security	Water Security	Migration
Domestic governance	Strong	Moderate	Strong to moderate	Weak	Workers: weak Refugees: weak
Regional institutions	Moderate	Weak	Weak	Moderate	Workers: weak Refugees: weak
Global institutions	Weak	Moderate	Strong	na	Workers: weak Refugees: moderate

tions, protocols, and policies that overcome the collective action problem and provide international public goods that meet transnational security threats. We ranked institutions strong if they were able to design and enforce compliance with agreements that produced global and regional public goods. We ranked institutions moderate if they were able to exchange information and reach some agreements on responding to transnational security issues, but were largely unable to compel compliance with these agreements. We ranked institutions weak if they had little capacity to exchange information and devise policies to confront transnational security threats, and had no enforcement mechanisms. In this concluding chapter, we look issue by issue at the efforts of states, regional actors, and global institutions to cooperate on these threats, and then offer an explanation for our findings.

Climate Change

Climate change in Southeast Asia is projected to have dramatic effects on the region. Sea-level rise will pose enormous problems for island nations and coastal areas; increased rainfall will cause flooding in Jakarta, Bangkok, and other massive urban areas unless significant effort and resources are increased to improve pumps, canals, and drainage systems. Droughts will test the ability of millions of small farmers to continue to feed themselves, and it will impact yields for large farmers who need to be able to grow enough to meet urban, regional, and global demand. In order for Southeast Asia to do its part to slow greenhouse gas emissions and thus slow climate change, there will need to be more robust efforts to curtail deforestation. Despite years of discussing ways to move forward with REDD+, which involves conservation, sustainable forest practices, and the enhancement of forest carbon stocks, countries in Southeast Asia have made only lackluster progress on protecting forestland.[2]

In Chapter 2 on climate change, we examined Indonesia's experiences in attempting to address deforestation. Indonesia has committed to reducing GHGs by 27 percent, placed a moratorium on the conversion of forestland, and invested significant time, money, and technical expertise to create an effective system to map and monitor forest activity. However, even with a large financial incentive of up to $1 billion offered from Norway for a verifiable reduction in emissions, Indonesia has failed to significantly reduce the legal and illegal logging and clear-cutting that produce them. Such cutting has included the burning of peatlands, which causes the horrific haze that regularly blankets not only Indonesia but also Singapore and Malaysia.[3] What has made success illusive on this problem?

First, powerful interests from logging companies, palm oil companies, mining companies, and other business interests have curried favor at both the local and national levels of government to prevent the implementation of national policies designed to curb deforestation. Second, bureaucratic reorganization and competition over forest use among government agencies such as the Forestry Ministry (now combined with the Environment Ministry), Ministry of Agriculture, and Ministry of Mining also hinder implementation, which illustrates the problems associated with horizontal cooperation. Third, poor compliance and cooperation between national-level government agencies and local officials and offices tasked with oversight and implementation is an example of the difficulties of promoting vertical cooperation. Finally, one of the most significant problems impeding greater success on curtailing deforestation is corruption. When business groups can buy favors or pay off inspectors, government officials, and judges so that their lack of compliance with laws and regulations are not monitored or enforced, then it is hard to see how deforestation will be ended.

Regional and international bodies are trying to help states address climate change problems and change behavior. The ASEAN Socio-Cultural Community has been tasked with developing ASEAN's strategy to address climate change. ASCC has stated that ASEAN will participate in global mechanisms to combat climate change and improve environmental sustainability. Such efforts include both mitigation and adaptation mechanisms. The ASEAN Economic Community also covers climate change, articulating that ASEAN countries should be working toward mitigating GHGs and broader measures to ameliorate climate change more generally.[4] ASEAN, however, has taken few concrete steps to get Southeast Asian states to move toward these goals, instead serving as a venue for discussions of best practices and a forum for the dissemination of scientific and technical knowledge from global actors. The ASEAN Agreement on Transboundary Haze Pollution is the most concrete treaty at the regional level, laying out actions necessary to prevent, monitor, and put out forest fires that cause the haze, but it lacks enforcement mechanisms and, as evidenced by the devastating haze of 2015, has little real impact on the ground. The lack of a multilateral compliance mechanism led Singapore to undertake unilateral legal action against Indonesian firms responsible for the 2015 haze in an effort to make those responsible for producing haze bear the social, economic, and health costs of the negative externalities they create. Regional efforts at promoting cooperation on climate change are weak because the country with the largest stake in the issue, Indonesia, to date has lacked the capacity to lead despite making commitments at the global level that indicate a desire to do so.

Because climate change is the issue we analyzed that most closely approximates a global public good, there is a robust set of institutions at the international level to respond to it. The UN Intergovernmental Panel on Climate Change provides comprehensive scientific studies on global environmental conditions that provide the basis for global discussions under the auspices of the Conference of Parties of the UN Framework Convention on Climate Change. Convened in Paris, the Conference of the Parties 2015 (COP15) agreed in principal to take concrete measures to limit and roll back emissions to keep global temperature rise below 2 degrees C. There are several noteworthy elements about this agreement. First, the United States and China, the world's two largest emitters, demonstrated a willingness to commit to the agreement. Although the Donald Trump administration has subsequently announced that the United States will begin the four-year process of pulling out of the Paris Agreement, its decision has been widely condemned, no other major country has followed suit, and over 150 countries have ratified the Paris Agreement, which all illustrate a shift in global attitudes regarding the necessity to make formal commitments to reduce greenhouse gases.[5] Second, for the first time, wealthy countries have agreed to provide financial assistance to developing countries to help those most vulnerable to climate change reign in their emissions, marking an important breakthrough in climate negotiations. Indonesia as a developing country is eligible for such assistance, illustrating how multilateral efforts such as the Paris Agreement and its bilateral agreement with Norway are examples of cooperation under the fragmented climate governance architecture. Despite these cooperative global efforts, regional and international institutional involvement has not been sufficient to date in curbing deforestation in Indonesia. In this case, what is necessary to reduce GHGs are domestic reforms within Indonesia to promote more effective forest governance. Whether the new policies the Jokowi administration adopted in 2016 to strengthen forest governance will be strong enough to significantly reduce GHGs remains to be seen.

Food Security

Problems of food and water security, like the problem with haze created by forest fires, are local and regional problems, but are also ones connected to larger global dimensions. Southeast Asia will need leadership and greater cooperation in the years to come if the region is to address food needs in a changing global and regional context. Climate change, urbanization, declining investment in agriculture, changing diets, and population increases will continue to stress food stocks. In Chapter 3 on food security, we examined the rice trade in Southeast Asia as an example of how countries increasingly must look at their food security as a transnational issue. In 2007, at the height of peak commodity prices, India's decision to use rice for its domestic food program meant that there was a sudden decrease in the amount of rice being traded on world markets. Thailand and Vietnam reacted by cutting their exports of rice, further escalating the problem. This triggered a massive price spike and, more importantly, hoarding of rice stocks throughout Southeast Asia. These were domestic political choices that had negative spillover effects, triggering a regional crisis. Food security is built on the pillars of the availability, access, and affordability of food. During the crisis, rice-importing countries such as the Philippines and Malaysia found themselves facing the problem of reduced access to food at affordable prices. Wealthy Malaysians were able to afford to purchase rice, but the Philippines was unable to provide for its citizens' food security and people actually went hungry. Only the decision by the United States, Japan, and the World Trade Organization to allow Japan to sell stockpiles of US rice sitting in Japanese warehouses calmed rice markets, reduced prices and scarcities, and thereby ended the panic.[6]

Chapter 3 demonstrates how the interests of countries that import rice and those that export rice are dramatically different, particularly in times of crisis. These divergent interests have come about in large part because government officials perceive food security as national security and therefore prioritize meeting their own citizens' food needs ahead of people living in other states. It is encouraging to see that governments equate the human security of their citizens with national security. But this also means that, when rice becomes scarce, exporting countries will adopt beggar-thy-neighbor policies by imposing restrictions on the amount of food sent abroad, with negative consequences for importing countries.

In Southeast Asia we have seen a two-pronged approach to transnational food security: first, the development of steps to avoid crises like that of 2007 in the rice market; and second, cooperative measures to stockpile commodities in the event that panics occur. Liberal economics would prescribe greater open markets as a cure for food scarcities. As Chapter 3 demonstrates, the rice market lacks transparency, and uncertainty about the available quantities of rice helped trigger the hoarding and price hikes that produced the panic. In theory, importing and exporting countries would both benefit from greater transparency and liberalization of the rice trade. There could be big payoffs from greater efficiency and better allocation of resources: more stability of prices and supplies would reduce the tendency of actors to hoard stocks and, therefore, the likelihood of price panics. While importing countries would benefit from freer markets in times of stable availability and prices, the importance of food security at home means that importing countries are working to ensure greater access to food by building superfarms or signing bilateral trade deals to lock in needed food imports.

In reality, however, exporting countries are not seeking freer markets for rice production. Chapter 3 demonstrates how domestic political calculations trump more cooperative efforts to coordinate food policy in times of need. Rice exporters like Thailand and Vietnam would much prefer to control where and how much of their rice crop is sold outside their borders. There are two reasons for this: first, they want to be able to prioritize their own citizens' food needs should it be necessary; and, second, they want to be able to reward or co-opt their own producers for political and economic purposes. One might expect that export firms would encourage their governments to push for more open trade policies to improve their market access, but that is not always what happens. Thailand's assured price guarantees were a political favor to farmers, but they hurt exporting firms as Thai rice is no longer as competitive on world markets.[7] And the high cost to the government has now meant that payments to farmers are long overdue, hurting the very group that the government wanted to help.[8]

The biggest exporters in Southeast Asia are Vietnam, Thailand, and Cambodia. None of these countries is interested in or willing to shoulder the burden of acting like a hegemon and taking the lead in promoting more transparent markets for rice. If these countries pushed together for greater market openness in rice and greater information sharing about rice crops, then all countries (importing and exporting alike) would feel less insecure about access to rice supplies and could make decisions about how to best assist their poorest citizens. Indonesia is the largest country in the region but, in the case of the rice trade, Indonesia has little power to get countries to agree to new rules and practices. Indonesia produces a large amount of rice, but it is not a major rice exporter and thus it would be hard for it to lead by example or to convince exporters that it had their interests in mind.

In addition to measures to prevent panics in the rice market, the second mechanism to promote food security consists of nonmarket efforts to stockpile or otherwise coordinate food distribution in times of crises. ASEAN has repeatedly pledged to embrace food security as a matter of permanent and high priority. It created and implemented the ASEAN Integrated Food Security Framework and the Strategic Plan of Action on Food Security in the ASEAN Region. These measures are overseen by the periodic Senior Officials Meetings of the ASEAN Ministers on Agriculture and Forestry. In the wake of the 2007–2008 rice crisis, the ministers also agreed on the creation of the ASEAN Plus Three Emergency Rice Reserve, and the ASEAN Food Security Information System, a permanent plan to meet emergency requirements for rice, and an early warning system to provide information on agricultural commodities.[9]

The thirteen signatory states will stock up 787,000 tons of rice to be used in case of emergency or instability in rice supplies.[10] This will help in some limited situations, but it would be even better to agree on more open mechanisms for trade and to agree on binding practices. These measures have not happened. The reason that we do not see enforcement mechanisms in agreements is because of resistance from exporting states. It is hard to see that Thailand, Cambodia, or Vietnam would be in a position to take the lead on forging greater regional cooperation on the rice trade. Thus, there is little impetus or momentum for greater institutional cooperation within Southeast Asia for regulating the rice trade. Instead, countries are focusing on domestic options for ensuring food security. It remains to be seen if states have been successful enough at these measures to mitigate the next crisis and meet increased demand over time. Ultimately, in the case of access to food, national governments view human security as national security, and they want to take steps to use their state power to guarantee a supply at an affordable price rather than rely on market mechanisms. It is hard to see how this view would change.

The complex or hybrid nature of the actors involved in trying to coordinate on food security—private producers, government procurement agencies, private import and export firms, and regional organizations, just to name a few—makes clear how difficult it is to get all parties to agree on solutions to transnational food crises. Where there has been successful cooperation, such as on APTERR, we see an example of how hybrid systems of governance can work. But this is one small and very specific example of a larger and more difficult puzzle over how government, and nongovernment, and global organizational actors can cooperate on shared problems. The broader problem of coming up with ways to make rice markets more transparent and resilient to crises has not been solved. Diverse, sometimes conflicting and fragmented, interests and dynamics are hard to overcome.

Water Security

Water security is an area where cooperation is, and increasingly will be, essential for human security in the region. Despite its lush tropic topography, Southeast Asia is actually water insecure. Pressures from rapid economic development, industrialization, urbanization, and climate change have put even greater stress on already fragile water supplies. Bangkok and Jakarta have suffered from devastating floods over the past five years, and now the Mekong River area is feeling the dire consequences of two years of record-breaking drought. The drought has led to dramatic drops in crop yields for rice, sugar, cassava, coffee, and other food products. And it is increasing salinity in the groundwater used for agricultural irrigation, thus ruining even greater acreages of crops in large areas of Vietnam.[11]

Our focus in Chapter 4 was a look at water security as a common problem and therefore we examined attempts to share the resources of the Mekong River. China, Myanmar, Laos, Cambodia, Thailand, and Vietnam all border the Mekong and it is the main source of water for drinking, bathing, transportation, and economic livelihoods (in the form of irrigation of crops, industry, and fishing) for millions of people living near it. "Around 80 percent of the Greater Mekong's 300 million people depend on healthy natural systems such as rivers, forests and wetlands for their food security, livelihoods and customs."[12] Ensuring water security to all the region's people requires states along the river to agree to guidelines and restrictions on how the river is used.

Chapter 4 demonstrated that domestic policy choices in riparian states are made based on powerful interest groups, rather than concern for the regional collective good and sustainable public access to the Mekong's water. The clearest example of this is the dam building under way in China and Laos. These countries construct dams to enhance their power generation capabilities at the expense of communities downstream. Chinese dam building on the Mekong is consistent with other large-scale hydropower projects in that country. Although the Chinese economy has slowed in recent years, power generation and energy resources more generally have been critical preoccupations in China as it races to further industrialize and bolster economic growth. In the case of Laos, powerful interests from Lao, Thai, and Chinese construction firms and utility companies have successfully influenced the Lao government to support and move ahead with dam construction. Construction firms will profit from building the dams, and energy companies hope to then sell the power to countries like Thailand and China. Few environmental impact assessments have been conducted, nor have comprehensive studies been undertaken to address the economic, social, and cultural impact to the many communities that will be displaced and adversely affected by changing water flows along the river. This is a perfect example of private and privileged interests being favored over the larger public.

As with other examples in this book, there are regional organizations that were created to help members address common interests in the Mekong's resources, yet these institutions have not been effective at addressing the most pressing problems of their members. The Mekong River Commission was created in 1995 to continue and enhance the work of the Mekong Committee. Members include Thailand, Cambodia, Laos, and Vietnam. China and Myanmar are observers. MRC has a broad list of issues in its bailiwick: drought and flood management, agricultural and irrigation needs, fisheries and the environment, hydropower, and navigation. At its core, MRC is supposed to ensure the reasonable and equitable utilization of the Mekong's water.[13] While MRC was useful in earlier efforts to map the river and improve surveys of the river's resources, it has been less useful in addressing the harder and more contentious issues of dam building, environmental stresses, and equitable use of the river. While Western donors had previously funded many of MRC's activities, several donors have cut back their support because of organizational mismanagement and MRC's inability to forge cooperative solutions on the issue of dam construction in Laos, which as noted above is favored by Laos, Thailand, and China but staunchly opposed by Cambodia and Vietnam.[14]

The Mekong originates in the Tibetan Plateau, inside Chinese territory. China is not a member of MRC and has made it clear that it does not have any interest in compromising or agreeing to water-sharing arrangements with any downstream neighbors. In fact, at the 2014 ASEAN-China summit meeting, Chinese premier Li Keqiang proposed the establishment of the Lancang-Mekong River Cooperation Mechanism. The organization came together quickly under Chinese leadership and encouragement. It ostensibly aims to promote cooperation and consensus on political security, economics, and sustainable development surrounding the river and its use. China views this as another way to facilitate and improve transportation networks and trade links to Southeast Asia. This enables China to ensure access to natural resources and power generation, and enhances China's ability to influence its neighbors. China portrays itself as a good neighbor; for example, when it released water in 2015–2016 to alleviate some of the worst effects of the drought. However, China continues to resist the idea of codifying rules for water use, dam construction, environmental guidelines, and so forth. And Laos and Thailand have been willing to go along with China to reap quick economic gains while other countries are more worried about Chinese power and encroachment on sovereignty and the impact of dam construction on their own citizens' use of the river. Vietnam is the most concerned with these issues.

Water security is an example of where there is a regional leader, China, who has the capacity to alter outcomes and ensure that regional public goods are provided that maximize human security for millions of Southeast Asians. China is taking the lead, but is not promoting legal norms and cooperation. Instead, it is steering countries toward decisions that benefit China and its allies rather than working toward an equitable, transparent, and sustainable sharing of the Mekong's resources. This could very well leave farmers in downstream countries much worse off. China's determination to act in its narrow national interest rather than the broader regional one is a classic case of beggar-thy-policies that will harm its neighbors. Chinese leadership in this instance is not as a hegemon providing public goods; instead, Chinese institution building and behavior is geared to providing China, and perhaps elites in Laos and Thailand, greater gains.

Avian Influenza

Our discussion of H5N1 in Chapter 5 demonstrates the fragmented governance architecture that exists between international organizations, national governments, bureaucratic agencies within governments, and those actors implementing policies at the local level. Fears of an outbreak of avian influenza, and the fear that if the virus mutates it could spread easily from person to person, has spurred the international community to play a greater role in public health oversight and delivery in Southeast Asia. Attempting to combat avian influenza shines a light on the internal power struggles among agencies and bureaucracies seeking resources and influence.

Difficulty in eradicating the incidence of H5N1 stems from weak government capacity, which is caused in part by a lack of money and poor infrastructure. Financial resources are needed to compensate farmers for culling their flocks, for monitoring, and to provide educational outreach. Weak public health systems and poor physical infrastructure, such as roads and information networks, hamper the implementation of effective policies. There also needs to be effective mechanisms for government health officials to carry out these programs at the local level. This requires good vertical cooperation between national and local-level officials. Vietnam and Thailand have been more successful at these measures than Indonesia and Cambodia.[15]

Cambodia and Indonesia have demonstrated that competition among government bureaucracies for resources has hindered efforts to educate, prevent, and treat H5N1 transmission. Choosing not to pay farmers to cull sick and potentially sick birds has also weakened the effectiveness of government campaigns. In contrast, Vietnam and Thailand have overcome these problems and taken more vigorous measures, including compensation for culling, to combat the disease. In these two cases, Vietnam's strong unitary state and Thailand's more professional bureaucracy and more experienced and extensive public health system made it easier for them to tackle avian influenza.[16] As a large exporter of poultry, Thailand had the added incentive to act quickly and decisively against H5N1 to protect a lucrative sector of the economy. It is telling, however, that Thailand's decision not to use vaccines in preventing H5N1 further bolsters the power of exporters' interests over those of small farmers who favored vaccination. Exporters resisted vaccination because it would have rendered their poultry ineligible to export to the European Union. On the specific issue of vaccination, this is an example of how private interests can prevail over the larger national interest. Nevertheless, in the Thai case more broadly, there was significant cooperation between private and public sector actors that significantly reduced outbreaks and transmission of avian influenza, which was the wider public good.

Regional and international efforts to combat H5N1 have been spearheaded by the WHO with some more modest efforts coming from ASEAN. A myriad of existing ASEAN bodies is already in operation to address the threat of pandemics, but ASEAN's difficulty in promoting cooperation lies with the wide capacity gaps in health systems among member states and perhaps the lack of willingness to create a regional fund to help poorer countries follow accepted protocols for culling diseased flocks.[17] In July 2012, the ASEAN Expert Group on Communicable Diseases developed a plan to deal with EIDs. The plan has two focus areas: (1) prevention and control of emerging infectious diseases; and (2) enhancing the regional supportive environment. The aims of the first focus area are to focus on prevention and control of EIDs by setting goals such as strengthening regional cooperative arrangements for integrated efforts at prevention, control, and preparedness for EIDs; developing regional support systems to narrow the gap among ASEAN member states in addressing EIDs; strengthening cooperation in tracing and health quarantine measures; strengthening surveillance measures for HIV/AIDS, malaria, tuberculosis, and so forth; promoting collaboration in research and development on new medicines; and ensuring stockpiles of antiviral medications and supplies for all member states for rapid response and containment of potential pandemic influenza. Details of the second focus area include: strengthening and maintaining regional support systems to narrow the gap in addressing EIDs; sharing of best practices in improving access to primary health care; strengthening regional clinical expertise through professional organizations; and information sharing to assist in response to EIDs as well as to other crises such as climate change and food security problems.[18]

Despite sounding like a comprehensive plan of action from ASEAN to deal with EIDs, the reality is much different. The funding is coming from the WHO and wealthy donor nations such as the United States. ASEAN seems to be making the most progress on the least controversial of these measures such as establishing websites to share information on EIDs (maintained by Indonesia). Countries have not always been forthcoming in sharing information, and compliance with best practices has been spotty at best. Lack of compliance stems not from lack of funds or information but rather from different preferences within countries on how to spend the money. In Indonesia, there is bureaucratic infighting and claims of sovereignty over who would control testing and vaccine development. This squabbling contributed to a much slower and weaker response than we should have seen. In Cambodia, corruption and the desire to use aid money to reward political supporters, rather than positively impact health outcomes, often dictated the choice of some programs over others so that little money was spent to compensate farmers or provide vaccines. Sadly, it is not surprising that too little has been done to cull sick or potentially sick flocks and cut down on disease transmission rates.

Migration

Finally, in Chapter 6, we took on two examples of transnational conflict relating to migration. Labor migration should be an area of cooperation among countries: in Malaysia and Singapore, labor is scarce and their economies benefit from employing economic migrants from countries with a labor surplus such as Indonesia and the Philippines. Despite signing bilateral Memorandums of Understanding to regulate migration, there is still a great deal of conflict between labor-sending and labor-receiving countries. Disagreement exists over issues such as changing economic conditions, working conditions, salaries, abuse, and criminal activities. Labor migrants who are welcomed or at least tolerated by receiving countries when economies are booming may be perceived as threats to local nationals when economic conditions deteriorate. Efforts by governments to protect their migrant workers from exploitative labor conditions in neighboring countries conflict with demands from employers in receiving countries to privilege their interests. Despite the opportunities for mutual gains, the reality is that there is little cooperation in the region over migration, and international protocols have had little effect on state behavior. Just as we found in Chapter 3 on food security that the divergent interests of rice exporters and importers hampered collaborative solutions that would leave both sides better off, with migration we see that different preferences of sending and receiving countries are an obstacle to cooperation. Indonesia has struggled to protect its citizens working abroad, particularly the women who leave Indonesia to go to Malaysia as domestic workers. Migrant workers have faced abuse in Malaysia and activists in Indonesia have increasingly pressured their own government to do more to safeguard their personal security.

Both the administration of Susilo Bambang Yudhoyono and now the government of Joko Widodo (aka Jokowi) have had a hard time addressing the issue with their Malaysian counterparts. In 2015, this led Jokowi to call for a moratorium on Indonesian women going abroad to work as housekeepers and nannies. If Indonesia could not protect its workers from threats to their well-being abroad, one solution was to prohibit them from working overseas in the first place. Many migrant advocates have criticized the administration for this decision, arguing that banning labor migration will have perverse consequences. If the needed number of jobs are not created at home, or other steps to alleviate poverty implemented, women will continue to travel abroad, only now they will do so outside of formal channels which puts them at greater risk. Nusron Wahid, head of BNP2TKI, has stated that with the Indonesian economy growing at 5 percent, only 1.5 million of the 2.8 million new entrants to the job market can secure work at home, which creates strong push factors for outward migration.[19] Eni Lesteri, chairwoman of the International Migrants Alliance, agrees with this concern stating that the "moratorium makes woman more vulnerable because they will migrate anyway."[20] Additionally, within Indonesia and Malaysia there are private actors who profit from migrants and put their personal pecuniary interests over the broader one of ensuring the safety of their follow citizens. Labor brokers in both countries would like to see migration continue since they profit from it, but their interests are at odds with political interests of elected officials who want to be seen doing something about the problem. The problem, however, is different in sending countries and receiving countries.

Labor migration has localized impacts and does not come close to approximating a global public good. As a result, the global efforts to govern migration are weak. The International Convention on the Protection of the Rights of All Migrant Workers and Members of Their Families lays out a standard for the treatment of migrant workers to guide policy, including rights to freedom of religion, a fair wage, and freedom from servitude and arbitrary determination.[21] Indonesia and the Philippines, two large sending countries in Southeast Asia have signed the convention, but Malaysia and Singapore, two primary labor importers have not. Similarly, efforts to strengthen cooperation at the regional level have resulted in a number of ASEAN declarations on this issue, but no binding agreements. The same divergence of interests that stymie cooperation at the international level hinder it at the regional level as well.

The second migration issue that we examined in Chapter 6 is the plight of the Rohingyas fleeing Myanmar and being refused asylum throughout the region. Rohingyas have been persecuted horribly within Myanmar and most are stateless. Violence has been perpetrated on the Rohingyas off and on since independence. As Myanmar has begun the process of political change, that violence has increased. Since 2012, anti-Muslim violence, state propaganda, and discriminatory laws have caused 140,000 Rohingyas to become internally displaced persons and another 120,000 to seek safety abroad.[22]

There are global agreements on refugees. The 1951 Convention Related to the Status of Refugees and its 1967 Protocol define a "refugee" as someone who has fled their country due to a well-founded fear of persecution, and as a result of this fear is unable to return to their home country. Rohingyas are clearly refugees and should be accorded protection offered by the convention. Signatories to the convention have an obligation to review their refugee application and, if recognized as such, provide for their well-being. The convention is a clear international effort to protect human security.

The Philippines is a signatory to the convention but the Southeast Asian states most affected by the Rohingyas, such as Thailand, Malaysia, and Indonesia, are not. As a result, they have no formal obligation to accept asylum seekers. Given their proximity to Myanmar, however, they are what the convention terms "countries of first asylum," or the places refugees seek protection temporarily and have their requests for asylum processed. UNHCR is the body charged with overseeing refuges, and has traditionally proscribed three main policies toward refugees: naturalization and resettlement in host countries, voluntary repatriation, and resettlement in third countries. Given growing opposition to accepting refugees in Western Europe and North America, the prospects for voluntary repatriation or resettlement in third countries are so slim that Southeast Asian countries fear being responsible for hosting the Rohingyas over the long term.

These fears led Southeast Asian states to adopt beggar-thy-neighbor policies in recent years as approximately 80,000 Rohingyas have fled from the Bay of Bengal since early 2014.[23] Thailand had become a hub for regional human smuggling activities. From 2012 to 2015, many Rohingyas paid smugglers to get them to Thailand, hoping then to travel overland to Malaysia or board other boats to Malaysia or Indonesia. In its 2014 "Trafficking in Persons Report," the US State Department downgraded Thailand to Tier 3, the bottom rank as a source, destination, and transit country for trafficking.[24] This designation and the discovery of mass graves in what was believed to be a smuggling detention camp led the Thai government to crack down on smuggling. Fearful of being caught, smugglers started abandoning their ships at sea.

In the spring and summer of 2015, an estimated 5,000–8,000 refugees from Myanmar were left stranded in open waters, when no Southeast Asian states would let them land.[25] Instead, the Malaysian, Thai, and Indonesian authorities pushed boats packed with Rohingyas back to sea after providing water and food. In 2016 and 2017, the situation in Myanmar has become worse and more Rohingyas are trying to leave. Clearly, if Myanmar's efforts to cleanse its country of Rohingyas can be viewed as creating a huge negative externality for its neighbors, then rejection of Rohingyas by Southeast Asian states indicates a strong desire to avoid paying the economic, social, and political costs of accepting them. The United Nations responded by pleading with Southeast Asian states to open their borders to prevent further suffering, and human rights activists slammed the not in my backyard policies.[26] Refusing to help refugees is not a new phenomenon, nor is it unique to Southeast Asia. In the 1970s, Southeast Asian nations first refused entry to those fleeing violence and persecution from the wars in Vietnam and Cambodia, but then agreed to take refugees temporarily as countries of first asylum because the United States and other states were willing to be countries of permanent asylum for these refugees. Without a third country being willing to take in Rohingya refugees, Malaysia, Thailand, and Indonesia have been reluctant to let Rohingyas into their countries. They have responded by attempting to pressure the Myanmar government to stem the flow of refugees and thereby reduce the threat, something the government of Myanmar has been unwilling to do. Cooperation on migration issues is weak and almost nonexistent. Without regional or international leadership on helping refugees or creating stronger protections for migrant workers, Southeast Asian countries have had trouble addressing these insecurities. The threats to human well-being do not accrue equally across state borders and national interests are widely divergent. Interests within countries sometimes clash as well.

Reasons for Poor Cooperation

Cooperation on nontraditional security threats would go a long way toward helping citizens and communities have greater access to food and water security, health care, protection from effects of climate change, access to labor markets, and safety from persecution. However, in looking at these challenges in Southeast Asia, we see little cooperation. There are a number of reasons why cooperation on transnational security issues in Southeast Asia is so weak. Our explanation stems from three major factors: first, competing internal interests and weak governance at the domestic level impede efforts of many governments to implement effective policies; second, there are competing national interests among regional states across a number of the transnational issues, particularly on migration, water, and food security; and, third, regional institutions in Southeast Asia are relatively weak.

Competing Domestic Interests and Weak Governance

Our first conclusion is that domestic governance hinders efforts to address these problems within national borders and makes it less likely that states will cooperate with each other across borders on common problems. Domestic political factors are at the crux of this discussion: Why is policymaking to provide public goods, such as adequate public health systems or well-regulated markets for food distribution, problematic? And why do these obstacles impede broader transnational coordination even though doing so would improve human security for a larger number of people? There are two broad sets of explanations for this: interest group politics, and problems of state capacity and corruption. We look briefly at each of these.

Often the explanation for beggar-thy-neighbor policy choices is that domestic political forces or key interest groups work hard to promote policies that benefit their interests at the expense of the larger collective good. For example, logging companies, mining firms, rice growers (or exporters), or power companies may be able to influence policies in ways that benefit their interests, even if these policies undermine efforts to combat climate change or address regional food and water supply issues. Also, even when states do agree in principle to a set of policies designed to promote the public good, and even if they do sign an international agreement to reduce greenhouse gas emissions, for example, enforcing these commitments is extremely difficult. Monitoring agencies may be weak and underfunded with little ability to offer carrots to induce compliance or sticks to compel private actors to change their behavior in line with the broad national interests rather than private ones. National governments may not have the capacity to enforce policies at the local level, which is indicative of the challenges of forging vertical cooperation in a fragmented system of governance. The ability of states to act in the broad national interests is also distorted by problems of corruption, where national or local-level officials are willing to turn a blind eye to policy violations in exchange for money or political support. Many countries in Southeast Asia have high levels of corruption. This makes policy enforcement difficult. Regardless of whether we are talking about measures to combat H5N1, where local officials may pocket money meant to be used to compensate farmers for culling infected flocks, or cases where logging firms use their political power, local patronage networks, and outright bribery to keep on cutting down forests as in Indonesia, corruption is a huge barrier to efforts to combat transnational threats. These domestic political dynamics occur throughout Southeast Asia and are a significant obstacle to better addressing the problems we analyze here.

Competing National Interests

Our second conclusion from the cases is that there are competing national interests on these nontraditional security threats that hamper cooperation. For example, the interests of exporters and importers of food are very different. Importing countries would like to see more open and transparent markets for trade in food commodities. Exporting countries must deal with competing interests of producers, trading firms, and their own population's food needs. Similarly, labor-sending and labor-receiving countries have distinct and competing interests. Indonesia wants to protect the rights of their workers overseas while Malaysia wants to protect the interests of their companies and employers, who often have little desire for greater regulation of working conditions and wages. In fact, Malaysia would like to be better able to turn the spigot on and off. That is, Malaysia would like to be able to import labor quickly and cheaply when the economy needs it, but be able to send workers home, or reduce the flow, when the economy slumps and domestic opinion turns antiforeign. The interests of countries such as Singapore and Malaysia, who have suffered from the haze resulting from peat fires in Indonesia, are different from the interests of powerful Indonesian forestry companies. Ultimately, it is Indonesia's responsibility to stop the uncontrolled burning and deforestation that also negatively impacts millions of its own citizens. On the Mekong River, the interests of power companies and construction firms in China and Thailand are different from the interests of small farmers in Vietnam and Cambodia who are impacted by the construction of new dams in Laos and China. These interests are competing rather than complementary. Thus, it is difficult to see how countries will find ways to reconcile them, both internally and at the regional level.

Weak Institutions and Lack of Leadership

Our third conclusion about why cooperation in Southeast Asia is weak stems from a lack of robust regional institutions. ASEAN, like the UN, is a broad general purpose organization that was originally created to address regional security and diplomatic issues, but has now expanded its agenda to cover transnational issues. Also like the UN, it is an intergovernmental organization composed of states that guard their sovereignty jealously. ASEAN decisionmaking requires a consensus vote by all ten members, which effectively gives each member veto power over the institution's decisions. Moreover, ASEAN lacks any effective enforcement capacity to uphold its decisions. The ASEAN Secretariat is weak and serves a coordination role, not a compliance one. Like the UN, therefore, ASEAN can provide a venue for the exchange of information and the crafting of cooperative solutions, but lacks mechanisms to overcome the collective action problem. Effective international institutions exist where there are common interests and where those interests are bolstered by the capacity of states to take and compel action. The divergent interests of Southeast Asian states make it difficult to envision a significant strengthening of regional institutions.

Efficacy of institutions also reflects leadership. In the UN, leadership is often provided by members of the Security Council, particularly the five permanent members including the United States and China. In ASEAN, there are no permanent members of a group similar to the Security Council that can provide leadership on transnational security issues. Indonesia is by far the largest ASEAN member and is often referred to as ASEAN's de facto leader or primus inter pares. Indeed, Indonesia often plays a leadership role on political and traditional security issues. On nontraditional issues, however, Indonesia lacks the credibility to lead due to its poor domestic governance. How can Indonesia promote effective collaboration solutions on combating the threat from global warming when it cannot prevent the haze? Indonesia certainly has an interest in cooperative solutions to EIDs, but its own domestic capacity is lacking. Moreover, during the avian influenza outbreak, Indonesia's health minister appeared more interested in revising the global protocols for virus sharing at the WHO level than in ASEAN efforts to promote cooperation on EIDs. Within ASEAN, there is no effective leadership in part due to the competing interests of member countries. ASEAN's weakness is exacerbated by the consensus decisionmaking that holds cooperation hostage to any one country.

At the broader regional and global levels, there simply is no single actor who has the broad interest and willingness to pay the costs of overcoming the collective action problem that hegemonic stability theory claims is often necessary for effective cooperation. Climate change is the transnational threat that most closely approximates a public good, but no one country can promote cooperation on its own. Instead, we do see a privileged group or a small number of countries with an interest in the issue attempting to lead. The decision by the world's two largest emitters of GHGs, China and the United States, to make commitments just ahead of the Paris COP15 talks to encourage others to follow suit is instructive. The Paris Agreement, however, has no enforcement mechanism, and there is no way to compel China and the United States to abide by their own commitments, as shown by the Trump administration's decision to pull out of the agreement.

Indeed, on only one transnational issue covered in this book is there a country with the capacity over the issue to play the role of hegemon: China on the Mekong. As the key upstream country, China has the influence over the water flows, and could have used its position to craft a cooperative solution for the benefit of all the riparian countries. Instead, China has acted not in the collective interests of all riparian countries, but in its own narrow national interests to build dams to promote its own water security at the expense of its neighbors.

China, as the region's ascendant power, has clearly indicated a desire to play a larger leadership role in the region on traditional political and economic issues, as evidenced by the creation of the Asian Infrastructure Investment Bank. However, on matters of food security, climate change, and the spread of diseases, China has a poor track record of addressing these issues at home. Thus, it is hard to imagine Chinese leadership on these same issues within Southeast Asia. Chinese leaders are far more interested in preserving access to natural resources and sources of food that will contribute to China's continued industrialization, and which Chinese Communist Party leaders assume will help them maintain power, than they are in conserving resources. Its omnipresent pollution problems, its status as a food importer, its dam building along the Mekong, and its weakness at addressing public health problems at home make China unable to serve as a credible leader on transnational security threats.

While the United States has signaled interest in and concern over food security issues and over EIDs, outside of ad hoc agreements and some aid contributions, the United States has not tried to play a major role in promoting specific long-lasting action on these problems. The United States is not a member of ASEAN, which makes it difficult to envision how it could play a more significant role within the context of Southeast Asian regional institutions. The United States is an important actor on the issues we have discussed in this book, from the involvement of the Centers for Disease Control in EIDs, to helping fund some REDD+

activities, to agreeing to allow Japan to use stockpiles of US rice to end price spikes and panics over rice shortages in 2008. However, these actions reflect US interests on specific issues, rather than a more comprehensive policy. US policy toward Southeast Asia is primarily focused on traditional security questions such as the rise of Chinese power, freedom of navigation through the region's strategic waterways, and protecting US economic interests. Nontraditional security issues, like the ones we have discussed, still take a back seat for the United States except in times of crisis. There seems to be little likelihood that either China or the United States will spearhead greater cooperative efforts within Southeast Asia to address these transnational security questions.

The best chance for greater cooperation will have to come from Southeast Asian countries themselves. Unfortunately, in Southeast Asia we see poor performance by states to adopt and implement policies to protect their citizens from transnational threats, and poor cooperation among states within regional and international organizations to facilitate cross-border coordination to solve common problems. But the conditions that impede cooperation are not set in stone. It is possible to imagine that domestic political conditions could change. Public pressure could change a government's calculation of risks and interests just as crises could lead to a new political will to take on powerful interests. Human security should be paramount to government leaders, and we look forward to a time when this will be the case in Southeast Asia.

Notes

1. Joe Cochrane, "Lives Still in Limbo, One Year After Southeast Asia Migrant Crisis," *New York Times,* June 20, 2016.

2. Food and Agriculture Organization (FAO), "Global Forest Resources Assessment 2015" (Rome: FAO, 2015), http://www.fao.org/3/a-i4808e.pdf.

3. Chris Lang, "Norway Admits that 'We Haven't Seen Actual Progress in Reducing Deforestation' in Indonesia," *REDD Monitor,* May 3, 2016. http://www.redd-monitor.org/2016/03/03/norway-admits-that-we-havent-seen-actual-progress-in-reducing-deforestation-in-indonesia/.

4. Bernadinus Steni, "Climate Change and ASEAN," *Jakarta Post,* May 10, 2011.

5. UN Framework Convention on Climate Change (UNFCCC), "Historic Paris Agreement on Climate Change," December 12, 2015, newsroom.unfccc.int/unfccc-newsroom/finale-cop21. It is useful to add here that the agreement by the United States and China in Paris built on an earlier 2014 bilateral agreement to roll back emissions of both countries. President Donald Trump announced that the United States will exit the accord; formally, this will occur over four years. In practicality, some US cities, states, and firms are already taking steps to comply with Paris and these efforts may continue despite Trump's actions. Hiroko Tabuchi and Henry Fountain, "Bucking Trump, These Cities, States and Companies Commit to Paris Accord," *New York Times,* June 1, 2017.

6. Tom Slayton and C. Peter Timmer, *CGD Notes: Japan and China Can Solve the Rice Crisis—But U.S. Leadership Is Needed* (Washington, DC: Center for Global Development, 2008).

7. This policy increased Thai rice supplies, which depressed both Thai and world prices, cut out private trading firms, and seems to have done little to ensure food security in the region in either the short term or longer term. And it has been so expensive for the government that it has found it difficult to meet its obligations to farmers.

8. Shuan Sim, "Thailand Rice Subsidy Scheme: What It Is and How It Toppled Thai Leader Yingluck Shinawatra," *International Business Times,* January 23, 2015.

9. Yang Razali Kassim, "ASEAN's Role in Asian Food Security," *Business Times*, September 14, 2011, 21.

10. Ibid.

11. Brahma Chellaney, "In Asia, Govts Struggle to Deal with Worsening Water Crisis," *Today Online,* April 28, 2016, http://www.todayonline.com/world/asia/asia-govts-struggle-deal-worsening-water-crisis.

12. World Wildlife Fund (WWF), "People of the Greater Mekong."

13. Mekong River Commission, "Mekong River Commission History," www.mrcmekong.org.

14. Luke Hunt, "Mekong River Commission Faces Radical Change," *The Diplomat,* January 22, 2016.

15. WHO, Regional Office for Southeast Asia, "Emerging Diseases: Avian Influenza" (WHO, Regional Office for Southeast Asia, 2015).

16. Thailand had improved its public health infrastructure in part to combat HIV/AIDS in the 1990s.

17. Gianna Gayle Amul, "ASEAN's Critical Infrastructure and Pandemic Preparedness," *NTSblog,* January 22, 2013, https://ntsblog.wordpress.com/?s=ASEAN's+Critical+Infrastructure+and+Pandemic+Preparedness&submit=.

18. Bounpheng Philavong, "ASEAN Regional Cooperation in Communicable Diseases and Pandemic Preparedness and Response," ASEF Network for Public Health, Thematic Working Group 1 on Regional Integration and Infectious Diseases, May 2006, https://www.asef.org/images/docs/1156-Presentation_4__ASEAN_Regional_Cooperation_in_Communicable_Diseases_and_Pandemic_Preparedness_and_Response_Dr_Bounpheng_Philavong.pdf.

19. Khoirul Amin, "Indonesia Seeks Greater Int'l Access for Its Workers," *Jakarta Post,* January 13, 2013.

20. Astrid Zweynert, "Indonesia's Maid Ban Increases Risk of Human Trafficking: Campaigner," Reuters, June 17, 2015.

21. UN General Assembly, *International Convention on the Protection of the Rights of All Migrant Workers and Members of Their Families,* Res. 45/158 (18 December 1990).

22. Council on Foreign Affairs (CFR), "Sectarian Violence in Myanmar," CFR Global Conflict Tracker, 2016, https://www.cfr.org/interactives/global-conflict-tracker#!/conflict/sectarian-violence-in-myanmar.

23. Global Centre for the Responsibility to Protect, "Populations at Risk: Myanmar," http://www.globalr2p.org/regions/myanmar_burma.

24. US Department of State, "Thailand: 2014 Trafficking in Persons Report," https://www.state.gov/j/tip/rls/tiprpt/countries/2014/226832.htm.

25. Eileen Ng, "Worries Grow for Rohingya, Bangladeshi Migrants as Malaysia, Indonesia, Refuse to Accept Them," *Canadian Press,* May 13, 2015.

26. Kanyakrit Vongkiatkajorn, "The Year Southeast Asia Faced Its Own Refugee Crisis," *Vice News,* December 27, 2015.

Acronyms and Abbreviations

AADMER	ASEAN Agreement on Disaster Management and Emergency Response
ACMW	ASEAN Committee to Implement the Declaration on the Protection and Promotion of the Rights of Migrant Workers
ADB	Asian Development Bank
AEC	ASEAN Economic Community
AEGCD	ASEAN Expert Group on Communicable Diseases
AFML	ASEAN Forum on Migrant Labour
AFSIS	ASEAN Food Security Information System
AHA Centre	ASEAN Coordinating Centre for Humanitarian Assistance on Disaster Management
AHMM	ASEAN Health Ministers Meeting
AICHR	ASEAN Intergovernmental Commission on Human Rights
AIFS	ASEAN Integrated Food Security Framework
AMAF	ASEAN Ministers on Agriculture and Forestry
AMBDC	ASEAN Mekong Basin Development Cooperation
APEC	Asia Pacific Economic Cooperation
APT	ASEAN Plus Three (China, Japan, and South Korea)
APTERR	ASEAN Plus Three Emergency Rice Reserve
ARF	ASEAN Regional Forum
ASCC	ASEAN Socio-Cultural Community
ASEAN	Association of Southeast Asian Nations
ASWGL	ASEAN Sectoral Working Group on Livestock
BIG	Badan Informasi Geospasial, Geospatial Information Agency
BNP2TKI	National Agency for the Placement and Protection of Indonesian Migrant Workers

BRG	Peatland Restoration Agency
CBAIC	Community-Based Avian Influenza Control Project
CDC	Centers for Disease Control (US)
CFCs	chlorofluorocarbons
CIAT	International Center for Tropical Agriculture
CIFOR	Center for International Forestry Research
CNMC	Cambodia National Mekong Committee
COP	Conference of the Parties
COP15	Conference of the Parties 2015
CPO Fund	Crude Palm Oil Supporting Fund
CPP	Cambodian People's Party
DAI	Development Alternatives, Inc.
DPR	Indonesian parliament
DSO	district surveillance officer
EAS	East Asia Summit
EGAT	Electricity Generating Authority of Thailand
EID	emerging infectious disease
EPA	Environmental Protection Agency
EU	European Union
FAMA	Federal Agricultural Marketing Authority
FAO	Food and Agriculture Organization
FSP	Food Security Policy
Gapki	Indonesian Palm Oil Producers Association
GATT	General Agreement on Tariffs and Trade
GDP	gross domestic product
GHGs	greenhouse gases
GISN	Global Influenza Surveillance Network
GISRS	Global Influenza Surveillance and Response System
GMS	Greater Mekong Subregion Economic Cooperation Program
GOARN	Global Outbreak Alert and Response Network
H5N1	strain of avian influenza
H7N9	strain of avian influenza
HPAI	highly pathogenic avian influenza
IDPs	internally displaced persons
IHR	International Health Regulations
ILO	International Labour Organization
IMF	International Monetary Fund
IMM	indicative moratorium map
IMMT	Institute of Minerals and Materials Technology
INGO	international nongovernmental organization
Inpres	Presidential Instruction (e.g., Inpres No. 10/2011)
IOM	International Organization for Migration

IPAPI	International Partnership on Avian and Pandemic Influenza
ISEAS	Institute of Southeast Asian Studies
IRRI	International Rice Research Institute
IWMI	International Water Management Institute
IWRM	Integrated Water Resources Management
KOPBUMI	Consortium of Indonesian Migrant Workers Advocacy
LMRC	Lancang-Mekong River Cooperation Mechanism
LOI	Letter of Intent
MAFF	Ministry of Agriculture, Forestry and Fisheries
MEM	Ministry of Energy and Mines
MEWR	Ministry of the Environment and Water Resources
MNC	multinational corporation
MOA	Ministry of Agriculture
MOH	Ministry of Health
MOMT	Ministry of Manpower and Transmigration
MOU	Memorandum of Understanding
MRB	Mekong River Basin
MRC	Mekong River Commission
MRV	monitoring, reporting, and verification
NAMRU-2	Naval Medical Research Unit No. 2
NAP	National Agricultural Policy
NEA	National Environment Agency
NGO	nongovernmental organization
NLD	National League for Democracy
NPR	National Public Radio
NTS	Centre for Non-Traditional Security Studies
ODA	official development assistance
ODI	overseas direct investment
OECD	Organisation for Economic Co-operation and Development
OIE	World Organisation for Animal Health
PDI-P	Indonesian Democratic Party of Struggle
PDSR	participatory disease surveillance and response
Perpres	Presidential Regulation
PHEIC	public health emergency of international concern
POLO	Philippine Overseas Labor Office
REDD	Reducing Emissions from Deforestation and Forest Degradation, became REDD+ after 2008
RELA	Ikatan Relawan Rakayat Malaysia
RSIS	S. Rajaratnam School of International Studies
SAFE	Strategies Against Flu Emergence
SARS	severe acute respiratory syndrome
SEANWFZ	Southeast Asia Nuclear Weapons Free Zone

SOM	Senior Officials Meeting
SOMHD	Senior Officials Meeting on Health Development
SPA-FS	Strategic Plan of Action on Food Security in the ASEAN Region
THPA	Transboundary Haze Pollution Act
TNC	transnational corporation
UN	United Nations
UNCLOS	UN Convention on the Law of the Sea
UNDP	UN Development Programme
UNECAFE	UN Economic Commission for Asia and the Far East
UNEP	UN Environment Programme
UNFCCC	UN Framework Convention on Climate Change
UNHCR	UN High Commissioner for Refugees
UNICEF	UN Children's Fund
UNIFEM	UN Development Fund for Woman
UNIPCC	UN Intergovernmental Panel on Climate Change
UPK4	Presidential Work Unit for Development Monitoring and Control
USAID	US Agency for International Development
USDA	US Department of Agriculture
WAHLI	Indonesian Forum for the Environment
WHA	World Health Assembly
WHO	World Health Organization
WTO	World Trade Organization
WWF	World Wildlife Fund

Bibliography

Acharya, Amitav. "The Future of Global Governance: Fragmentation May Be Inevitable and Creative." *Global Governance* 22, no. 4. (2016): 453–460.

Adamrah, Mustaqim. "RI Risks Businesses, Forests with US$1b Forest Moratorium Delay." *Jakarta Post,* January 6, 2011.

Adler, Emanuel, and Steven Bernstein. "Knowledge in Power: The Epistemic Construction of Global Governance." In *Power and Global Governance,* edited by Michael Barnett and Raymond Duvall, 294–318. New York: Cambridge University Press, 2005.

Adnan, Hanim. "Food Security Policy May Be Extended—Business News." *The Star Online,* April 10, 2011. http://www.thestar.com.my/business/business-news/2011/04/09/food-security-policy-may-be-extended/.

"Agreement on the Cooperation for the Sustainable Development," Article 5. Mekong River Commission, April 5, 1995.

Aguiba, Melody. "Philippines to Play a Critical Role in Global Food Security with 2012 Golden Rice Launch." *Jakarta Globe,* March 19, 2011.

Alagappa, Mutiah. *Asian Security Practice.* Stanford: Stanford University Press, 1998.

Albert, Eleanor. "The Rohingya Migrant Crisis." Washington, D.C.: Council on Foreign Relations, January 12, 2017.

Alexandra, Lina. "Indonesia Emerging." *The Diplomat,* August 11, 2011.

Amin, Khoirul. "Indonesia Seeks Greater Int'l Access for Its Workers." *Jakarta Post,* January 13, 2016.

Amul, Gianna Gayle. "ASEAN's Critical Infrastructure and Pandemic Preparedness." Health and Human Security by *NTSblog,* January 22, 2013. www.ntsblog.sg.net.

Anindhitya. "ASEAN Response to Combat Avian Influenza by ASEAN Secretariat, One Vision One Identity One Community." Jakarta: Association of Southeast Asian Nations (ASEAN), April 1, 2006.

Anjaiah, Veeramalla. "RI's Migrant Workers Remit Record $8.55 Billion in 2014: WB." *Jakarta Post,* April 20, 2015.

"Approaches to Controlling, Preventing and Eliminating H5N1 Highly Pathogenic Avian Influenza in Endemic Countries." Rome: Food and Agriculture Organization (FAO), 2011.

Armstrong, Paul. "Q&A: What's Behind Sectarian Violence in Myanmar?" CNN, October 26, 2012.

Arshad, Arlina. "Jakarta to Sue 9 More Firms over Forest Fires Linked to Haze." *The Straits Times,* March 1, 2016.

"ASEAN Response to Combat Avian Influenza." Jakarta: ASEAN Secretariat, April 2006.

"Asian Water Development Outlook: Measuring Water Security in Asia and the Pacific." Los Baños, Philippines: Asian Development Bank, 2013.

Audjarint, Wasamon. "ASEAN Struggles to Protect Laborers." *Jakarta Post,* May 30, 2016.

Aurora, Leony. "Indonesia Sets Aside 45% of Forest-Rich Kalimantan to Be World's Lungs." *Forests News Blog,* Center for International Forestry Research (CIFOR), January 24, 2012. https://forestsnews.cifor.org/7144/indonesia-sets-aside-45-of-forest-rich-kalimantan-to-be-worlds-lungs?fnl=en

Avery, Emma Chanlett. *International Efforts to Control the Spread of the Avian Influenza (H5N1) Virus: Affected Countries' Responses.* Report for Congress. Washington, DC: Congressional Research Service, August 24, 2006.

"Avian and Other Zoonotic Influenza." Geneva: World Health Organization (WHO), November 2016.

Babel, Mukand S., and Shahriar M. Wahid. "Freshwater Under Threat: Southeast Asia." Nairobi: UN Environmental Programme (UNEP), March 2009.

Bachtiar, Palmira Permata. "Chaotic Statistics of Indonesian Migrant Workers." *Jakarta Post,* January 26, 2012.

Backer, Ellen Bruzelius. "Research Paper Insights: The Mekong River Commission: Does It Work, and How Does the Mekong Basin's Geography Influence Its Effectiveness?" Lysaker, Norway: Fridjof Nansen Institute, 2007.

Barbier, Edward B. "Cash Crops, Food Crops, and Sustainability: The Case of Indonesia." *World Development* 17, no. 6 (1989): 879–895.

———. "Cash Crops, Food Crops, and Agricultural Sustainability." International Institute for Environmental Development, Swedish International Development Authority, Gatekeeper Series No. SA2. Sustainable Agriculture Programme, February 18, 2015. http://m.mekonginfo.org/assets/midocs/0002781-farming-cash-crops-food-crops-and-agricultural-sustainability.pdf

Barnett, Michael, and Steven Bernstein. "Power in Global Governance." In *Power in Global Governance,* edited by Michael Barnett and Raymond Duvall, chap. 1. New York: Cambridge University Press, 2005.

Barnett, Michael, and Martha Finnemore. "The Power of Liberal International Organizations." In *Power in Global Governance,* edited by Michael Barnett and Raymond Duvall, 161–184. New York: Cambridge University Press, 2005.

Baskoro, Faisal Maliki. "Nestle Shows Its Sweet Side to Greenpeace, Sinar Mas in Indonesian Palm Oil Fight." *Jakarta Globe,* May 17, 2010.

Biba, Sebastian. "China Drives Water Cooperation with Mekong Countries." *The Third Pole,* February 1, 2016.

Biermann, Frank, Philipp Pattberg, Harro van Asselt, and Fariborz Zelli. "The Fragmentation of Global Governance Architectures: A Framework for Analysis." *Global Environmental Politics* 9, no. 4 (2009): 14–40.

Bird, Jeremy, Wouter Lincklaen Arriens, and Dennis Von Custodio. "Water Rights and Water Allocation: Issues and Challenges for Asia." Los Baños, Philippines: Asian Development Bank, 2009.

Brandon, John. "Asia: The World's Most Water-Stressed Continent." Washington, DC: Asia Foundation, March 21, 2012.

Brown, Michael E., Owen R. Cote, Sean M. Lynn-Jones, and Steven E. Miller, eds. *New Global Dangers.* Cambridge: MIT Press, 2004.

Budiari, Indra. "Domestic Workers' Conditions at Home No Less Worrying." *Jakarta Post,* February 16, 2015.

"Burma: The Rohingya Muslims Face a Humanitarian Crisis." New York: Human Rights Watch, March 26, 2013.

"Burma, the Rohingya Muslims: Ending a Cycle of Exodus?" New York: Human Rights Watch Asia, September 1996.

Butler, Colin D. "Food Security in the Asia-Pacific: Climate Change, Phosphorus, Ozone and Other Environmental Challenges." *Asia Pacific Journal of Clinical Nutrition* 18, no. 4 (2009): 590–597.

———. "Indonesia Grants Exemption from Logging Moratorium for 3.6m Ha of Forest." *Conservation News,* December 21, 2011.

Butler, Rhett A. "50+ Companies Being Investigated or Punished for Indonesian Haze Crisis." Mongabay Series: Indonesian Forests. December 30, 2015.

Caballero-Anthony, Mely, and Alistair D. B. Cook, eds. *Non-traditional Security in Asia: Issues, Challenges and Framework for Action*. Singapore: Institute of Southeast Asian Studies (ISEAS), 2013.

Chan, Francis. "Singapore Targeting Private Firms Linked to Indonesian Fires." *Jakarta Post,* June 16, 2016.

Charles, Dan. "How Fear Drove World Rice Markets Insane." National Public Radio (NPR), November 2, 2011.

Chase, Robert S., Emily B. Hill, and Paul Kennedy, eds. *The Pivotal States: A New Framework for U.S. Policy in the Developing World.* New York: Norton, 1999.

Chellaney, Brahma. "China's Water Hegemony in Asia." *Washington Times,* April 26, 2016.

———. "In Asia, Governments Struggle to Deal with a Worsening Water Crisis." *Today Online,* April 28, 2016. http://www.todayonline.com/world/asia/asia-govts-struggle-deal-worsening-water-crisis.

"China Drives Water Cooperation with Mekong Countries." *The Third Pole,* February 1, 2016.

Chunsuttiwat, Supamit. "Response to Avian Influenza and Preparedness for Pandemic Influenza: Thailand's Experience." *Respirology* 13, no. 1 (2008): S36–S40.

"Climate Change: Basic Information." Overviews and Factsheets. Washington, DC: Environmental Protection Agency (EPA), January 17, 2017.

Cochrane, Joe. "Lives Still in Limbo, One Year After Southeast Asia Migrant Crisis." *New York Times,* June 20, 2016.

Commission on Global Governance. *Our Global Neighborhood: The Report of the Commission on Global Governance*. New York: Oxford University Press, 1995.

"Convention on Biological Diversity, Text of the Convention, Article 2 Use of Terms." https://www.cbd.int/convention/articles/default.shtml?a=cbd-02.

Corden, W. Max. "American Decline and the End of Hegemony." *SAIS Review* 10, no. 2 (2012): 13–26.

Curley, Melissa G., and Jonathan Herington. "The Securitisation of Avian Influenza: International Discourses and Domestic Politics in Asia." *Review of International Studies* 37, no. 1 (2011): 141–166.

"Current Forest Fires in Indonesia: Land-Use Fires and Smoke Pollution in Indonesia." Global Fire Monitoring Center, November 18, 2015. http://www.fire.uni-freiburg.de/GFMCnew/2015/11/20151019_id.htm#top.

Danh Tuyen, Le. "Food in Health Security in South East Asia." *Asia Pacific Journal of Clinical Nutrition* 18, no. 4 (2009): 493–497.

Daniel, Jessica R., Sandra Lee Pinel, and Jase Brooks. "Overcoming Barriers to Collaborate Transboundary Water Governance: Identifying Local Strategies in a

Fragmented Governance Setting in the United States." *Mountain Research and Development* 33, no. 3. (2013): 215–224.

"The Dark Side of Transition: Violence Against Muslims in Myanmar." Asia Report No. 251. International Crisis Group, October 1, 2013.

Davies, Michael, and Richard Woodward. *International Organizations.* Northampton, MA: Edward Elgar, 2014.

Dawe, Dave C. "The Philippines Imports Rice Because It Is an Island Nation." In *Why Does the Philippines Import Rice? Meeting the Challenge of Trade Liberalization,* ed. David C. Dawe, Piedad F. Moya, and Cheryll B. Casiwan. Los Baños, Philippines: International Rice Research Institute and the Philippine Rice Research Institute, 2006.

"Defining the Geopolitics of a Thirsty World: The Politics of the Mekong River." Austin: Stratfor, April 11, 2016. http://www.waterpolitics.com/2016/04/11/the-politics-of-the-mekong-river/.

Deibel, Terry. *Foreign Affairs Strategy: Logic for American Statecraft.* Cambridge: Cambridge University Press, 2014.

Dermawan, A., E. Petkova, A. C. Sinaga, M. Muhajir, and Y. Indriatmoko. "Preventing the Risks of Corruption in REDD+ in Indonesia." Bogor, Indonesia: Center for International Forestry Research, 2011.

Development Alternatives, Inc. (DAI). "Indonesia—Community-Based Avian Influenza Control (CBAIC) Project," accessed October 24, 2017. https://www.dai.com/our-work/projects/indonesia-community-based-avian-influenza-control-cbaic-project.

Dewan, Angela. "New Tech, Better Map on Tap to Protect Indonesian Forests." *Forests News Blog,* CIFOR, May 19, 2014. http://blog.cifor.org/22534/new-tech-better-map-on-tap-to-protect-indonesian-forests?fnl=en.

Dupont, Alan. *East Asia Imperiled: Transnational Challenges to Security.* New York: Cambridge University Press, 2001.

Dursin, Kanis. "ASEAN Agreement on Haze? As Clear as Smoke." Inter Press Service, October 10, 2015.

Ear, Sophal. "Avian Influenza: The Political Economy of Disease Control in Cambodia." *Politics and the Life Sciences* 30, no. 2 (2011): 2–19.

———. "Cambodia's Patient Zero: The Political Economy of Foreign Aid and Avian Influenza." Working Paper 398. Stanford: Stanford University Center for International Development, September 30, 2009.

———. "Emerging Infectious Disease Surveillance in Southeast Asia: Cambodia, Indonesia, and the US Naval Area Medical Research Unit 2." *Asian Security* 8, no. 2 (2012): 164–187.

Ear, Sophal, and Ian Scoones. *Avian Influenza: Science, Policy and Politics.* London: Routledge, 2010.

Emmers, Ralf. "Comprehensive Security and Resilience in Southeast Asia: ASEAN's Approach to Terrorism." *Pacific Review* 22, no. 2 (2009): 159–177.

Ewert, Sierra. "Eating the Last Drop: Changing Diets in Asia Challenge Future Water Security." Washington, DC: Asia Foundation, March 21, 2012.

F.,T. "Lies, Dams and Statistics." *The Economist,* July 26, 2012. https://www.economist.com/blogs/banyan/2012/07/mekong-river.

FAO Statistical Yearbook. Rome: FAO, 2007–2013.

Fidler, David P. "The Challenges of Global Health Governance." Working Paper. New York: Council on Foreign Relations, May 2010.

"Five Features of Lancang-Mekong River Cooperation." Beijing: Ministry of Foreign Affairs, People's Republic of China, March 17, 2016.

Flores, Jamil Maidan. "The Rohingya Need a Lifeline." *Jakarta Globe,* May 25, 2015.

Fogarty, David. "How Indonesia Crippled Its Own Climate Scheme." Reuters, August 16, 2011.

"Food Security and Famine Glossary of Terms." World Health Organization, 2012. www.who.int/trade/glossary.

Forster, Paul. "The Political Economy of Avian Influenza in Indonesia." Pathways to Sustainability. Sussex: Institute of Development Studies, Social, Technological, and Environmental Pathways to Sustainability (STEPS) Centre, March 10, 2009.

Fox, Jefferson, Yayoi Fujita, Dimbab Ngidang, Nancy Peluso, Lesley Potter, Niken Sakuntaladewi, Janet Sturgeon, and David Thomas. "Policies, Political-Economy, and Swidden in Southeast Asia." *Human Ecology* 37, no. 3 (2009): 305–322.

Freedman, Amy L. *Political Change and Consolidation: Democracy's Rocky Road in Thailand, Indonesia, South Korea, and Malaysia*. New York: Palgrave Macmillan, 2006.

Galvez, James Konstantin. "Philippines, Cambodia to Sign Rice Trade Deal." *Manila Times,* May 3, 2012.

Gertler, Isabel. "A Focus on Factors Affecting Avian Influenza in Indonesia." Sackville, New Brunswick, Canada: Atlantic International Studies Organization, December 2009.

"Global Conflict Tracker." Council on Foreign Relations. www.cfr.org/interactives/global-conflict-tracker#!/global-conflict-tracker.

"Global Forest Resources Assessment 2015." Rome: FAO, 2015.

"Global Forest Resources Assessment 2010. Country Report: Malaysia." Rome: FAO, 2010.

"The Global Framework for the Progressive Control of Transboundary Animal Diseases (TADs)." Paris: World Organisation for Animal Health (OIE), May 24, 2004.

"Global Health Security—Why It Matters." Atlanta: Centers for Disease Control (CDC), February 13, 2014.

"Global Influenza Programme." World Health Organization. http://www.who.int/influenza/en/.

"Greater Mekong: People and Communities." World Wildlife Fund. https://www.worldwildlife.org/places/greater-mekong#people.

Gunawan, T. Sima. "Questions Loom over 1-Billion Tree Planting Program." *Jakarta Post,* June 5, 2010.

Gupa, Aarit, Till Pistorius, and Marjanneke J. Vijge. "Managing Fragmentation in Global Environmental Governance: The REDD+ Partnership as Bridge Organization." *International Environmental Agreements* 16, no. 3 (2016): 355–374.

Gurr, Ted Robert. *Why Men Rebel*. Princeton: Princeton University Press, 1970.

Gutiérrez, Ramona Alikiiteaga, Monica Jane Naughtin, Srey Viseth Horm, Sorn San, and Philippe Buchy. "A (H5N1) Virus Evolution in South East Asia." *Viruses* 1, no. 3 (2009): 335–361.

Hameiri, Shahar. "Avian Influenza, 'Viral Sovereignty,' and the Politics of Health Security in Indonesia." *Pacific Review* 27, no. 3 (2014): 333–356.

Hameiri, Shahar, and Lee Jones. *Governing Borderless Threats: Non-Traditional Security and the Politics of State Transformation*. London: Cambridge University Press, 2015.

Hardum, Edi. "Legislators Seek End to Ban on Migrant Workers Going to Middle East." *Jakarta Globe,* September 3, 2015.

Hermesauto. "Singapore GDP Will Take Hit from Haze as Countries Issue Travel Warnings." *The Straits Times,* October 7, 2015.

"Highly Pathogenic Asian Avian Influenza A (H5N1) in People." June 11, 2015. Atlanta: CDC.

"Historic Paris Agreement on Climate Change." UN Framework Convention on Climate Change (UNFCCC), December 12, 2015. http://newsroom.unfccc.int/unfccc-newsroom/finale-cop21/.

Hoadley, Stephen. "The Evolution of Security Thinking: An Overview." In *Asian Security Reassessed,* edited by Stephen Hoadley and Jürgen Rüland, 3–36. Singapore: ISEAS, 2006.

Holodny, Elena. "Vietnam Is Being Crippled by Its Worst Drought in Nearly a Century." *Business Insider,* April 27, 2016.

Hossain, Faisal, and Katiyar, Nitin. "Improving Flooding Forecasting in International River Basins." *EOS, Transactions, American Geophysical Union* 87, no. 5 (2006): 49–60.

"How to Move Floodwater Through Bangkok." *IRINNews,* June 19, 2012.

Human Development Report 1994. New York: Oxford University Press for the UN Development Programme (UNDP), 1994.

Human Rights in ASEAN Online Platform. "ASEAN Committee on Migrant Workers." http://humanrightsinasean.info/asean-committee-migrant-workers/about.html.

Human Rights Watch. "Help Wanted: Abuses Against Female Migrant Domestic Workers in Indonesia and Malaysia." July 21, 2004. https://www.hrw.org/report/2004/07/21/help-wanted/abuses-against-female-migrant-domestic-workers-indonesia-and-malaysia.

"Human Security: Paradigm Shift or Hot Air?" *International Security* 26, no. 2 (2001): 87–102.

Hunt, Luke. "Drought Fans Tensions Along the Mekong." *The Diplomat,* March 12, 2016.

———. "Mekong River Commission Faces Radical Change." *The Diplomat,* January 22, 2016.

Hunt, Michael H. "American Decline and the Great Debate: A Historical Perspective." *SAIS Review* 10, no. 2 (2012): 27–40.

"The Importance of Biosecurity in Reducing HPAI Risk on Farms and in Markets." Paper prepared for the FAO International Ministerial Conference on Avian and Pandemic Influenza, New Delhi, December 2007.

"Indonesia—Community-Based Avian Influenza Control (CBAIC) Project." DAI. https://www.dai.com/our-work/projects/indonesia-community-based-avian-influenza-control-cbaic-project.

"Indonesia Bans Navy Medical Research Unit." *Navy Times,* April 10, 2008.

"Indonesia Cannot Afford to Ignore Food Security." *Jakarta Globe,* January 8, 2011.

"Indonesia Delaying Deforestation Ban." *Asia Sentinel,* April 14, 2011.

"Indonesia Files Complaint with Malaysia over Vacuum Cleaner Commercial." *Jakarta Globe,* February 4, 2015.

"Indonesia's Fire and Haze Crisis." Washington, DC: World Bank, December 1, 2015.

"Indonesia and Malaysia Agree to Offer 7,000 Migrants Temporary Shelter." *The Guardian,* May 20, 2015.

"Indonesia and Malaysia's Love-Hate Relationship." *Asian Correspondent,* April 8, 2014.

"Indonesia Palm Oil Industry 2015: Market Drivers, Challenges and Competitive Landscape Analysis." MarketersMedia press release, January 20, 2016.

"Indonesia's Palm Oil Permit Moratorium to Last Five Years." *Conservation News,* July 18, 2016.

"Indonesia Seeks Greater Access for Its Workers." *Jakarta Post,* January 13, 2016.

"Indonesia's Vice-President Jusuf Kalla Criticises Neighbours for Grumbling About Haze." *The Straits Times,* March 5, 2015.

"Indonesian Rice Production and Trade Policy." AsiaOne, March 3, 2011.
"International: Biofuels Spark Food-Versus-Fuel Debate." *Oxford Analytical Daily Brief,* December 7, 2007.
International Convention on the Protection of the Rights of All Migrant Workers and Members of Their Families. UN General Assembly, Resolution 45/158, December 18, 1990.
Islam, M. Rezaul. "Partnership Research Between Malaysia and Indonesia: A Case of Labour Welfare." *Social Research Reports,* no. 25 (2013): 7–20.
Jacobson, Phillip. "Malaysian Palm Oil Companies Say Their Concession Maps Are State Secrets." Mongabay, May 6, 2016. https://news.mongabay.com/2016/05/malaysian-palm-oil-companies-say-concession-maps-state-secrets.
Jailani, Abdulkadir. "The Need for an ASEAN Treaty on Migrant Workers." *The Diplomat,* January 22, 2015.
"Joint Press Statement of the Tenth Meeting of the ASEAN Ministers on Agriculture and Forestry and the Ministries of Agriculture of the People's Republic of China, Japan, and the Republic of Korea." Tenth ASEAN Ministers on Agriculture and Forestry (AMAF) Plus Three, Phnom Penh, October 24, 2010.
"Jokowi's Ban on Migrant Workers Unconstitutional: NGO." *Jakarta Post,* February 15, 2015.
"Jokowi, Najib to Appoint Envoys on Boundary Rows." *Jakarta Post,* February 7, 2015.
Jong, Hans Nicholas. "Govt to Put Moratorium on Firms with Idle Concessions." *Jakarta Post,* June 18, 2016.
Jotzo, Frank. "Indonesia's Role in International Climate Change Policy." *East Asia Forum,* October 11, 2011.
Jyotsna. "USDA Post Estimates 2015–16 Malaysia Rice Imports to Increase 5% Y/y to 1.05 Million Tons." *Oryza,* March 3, 2016.
Kallhoff, Angela. *Why Democracy Needs Public Goods.* New York: Lexington Books, 2011.
Karns, Margaret P., and Karen A. Mingst. *International Organizations: The Politics and Processes of Global Governance,* 2nd ed. Boulder: Lynne Rienner, 2010.
Kassim, Yang Razali. "ASEAN's Role in Asian Food Security." *Business Times,* September 14, 2011.
———. "The 19th ASEAN Summit: Tackling Floods, Food and Stability—Analysis." *Eurasia Review,* November 17, 2011.
Keo, Sovannarith. "Transboundary Water Conflict: A Case Study of the Mekong River Basin." Academia.edu.
Keohane, Robert, Peter Haas, and Mark Levy, eds. *Institutions for the Earth: Sources of Effective International Environmental Protection.* Cambridge: MIT Press, 1993.
Keohane, Robert O. *After Hegemony.* Princeton: Princeton University Press, 2004.
"Key Migration Terms." International Organization for Migration (IOM), 2016. www.iom.int/key-migration-terms.
Kindleberger, Charles P. *The World in Depression, 1929–1939.* Berkeley: University of California Press, 1973.
Kolesnikov-Jessup, Sonia. "Singapore Looks to China for Food Security." *New York Times,* September 27, 2010.
Korves, Ross. "Thailand's Domestic Rice Policies Upset Trade Patterns." Global Farmer Network, February 9, 2012.
Kovacevic, Michelle. "Norwegian Minister for Environment Praises Indonesia's Fight Against Climate Change." *Forests News Blog,* CIFOR, September 27, 2011. http://blog.cifor.org/4260/norweigian-minister-for-environment-praises-indonesias-fight-against-climate-change/.

Kunmakara, May. "Phnom Penh Autonomous Port to Push Rice Exports." *Phnom Penh Post,* June 8, 2012.
Kurniawan, Nanang Indra. "Getting the One Map Policy Right." Indonesia at Melbourne, May 11, 2016. http://indonesiaatmelbourne.unimelb.edu.au/getting-one-map-policy.
International Organization for Migration (IOM). *Labour Migration from Indonesia: An Overview of Indonesian Migration to Selected Destinations in Asia and the Middle East.* Jakarta: IOM, 2010. www.iom.int/cms/en/sites/iom/home/where-we-work/asia-and-the-pacific/indonesia.html. "
Lamb, Kate. "Indonesia's Fires Labelled a 'Crime Against Humanity' as 500,000 Suffer." *The Guardian,* October 26, 2015.
Lang, Chris. "Norway Admits that We Haven't Seen Actual Progress in Reducing Deforestation in Indonesia." *REDD Monitor,* May 3, 2016.
Larson, Christina. "Mekong Megadrought Erodes Food Security." *Science,* April 6, 2016.
"Law, International Water: The Customary International Law of International Waters." *Water Encyclopedia.* http://www.waterencyclopedia.com/La-Mi/Law-International-Water.html.
Liow, Joseph. "Malaysia's Illegal Indonesian Migrant Labour Problem: In Search of Solutions." *Contemporary Southeast Asia* 25, no. 1 (2003): 44–64.
Lloyd, Graham. "Indonesia Savours Fruits of the Forest." *The Australian,* September 30, 2011.
"Low Rain Forces Thai Rice Farmers to Consider Growing Other Crops Instead." *Japan Times,* April 8, 2016.
Mackenzie, John S., Patrick Drury, Ray R. Arthur, Michael J. Ryan, Thomas Grein, Raphael Slattery, Sameera Suri, Christine Tiffany Domingo, and Armand Bejtullahu. "The Global Outbreak Alert and Response Network." *Global Public Health* 9, no. 9 (2014): 1023–1039.
Maierbrugger, Arno. "Indonesia to Impose Ban on Maids Working Overseas Earlier than Expected." ASEAN Trade Union Council, February 21, 2015.
Makara. "Phnom Penh Autonomous Port to Push Rice Exports." *Phnom Penh Post,* June 8, 2012.
"Malaysia Country Summary." Human Rights Watch, January 2006. www.hrw.org/legacy/wr2k6/pdf/malaysia.pdf.
Mammadov, Kanan. "Review of 'The Power of Liberal International Organizations' by Michael Barnett and Martha Finnemore." In *Power in Global Governance,* edited by Michael Barnett and Raymond Duvall, 161–184. New York: Cambridge University Press, 2004.
Marchand, Alexandre. "Inside the 969 Movement." *Ruom,* June 8, 2013.
McKirdy, Euan. "Southeast Asia's Haze: 'A Crime Against Humanity.'" CNN, October 29, 2015, http://www.cnn.com/2015/10/29/asia/southeast-asia-haze-crisis/index.html.
"Mekong River Commission for Sustainable Development." Vientiane, Laos: Mekong River Commission, 1995.
"Mekong River Commission History." Vientiane, Laos: Mekong River Commission, 1995. www.mrcmekong.org.
"Migrant Workers in ASEAN: The Hidden and Neglected Workforce." ASEAN Trade Union Council, May 26, 2015.
"Migrant Workers' Rights to Social Protection in ASEAN." Munich: Friedrich-Ebert-Stiftung Regional Programme Asia, November 1, 2011.
Milner, Helen V. "International Theories of Cooperation Among States." *World Politic*s 44, no. 3 (1992): 466–496.

———. "International Political Economy: Beyond Hegemonic Stability." *Foreign Policy,* no. 110 (1998): 112–123.

"Ministry of Water Resources and Meteorology." Government of Cambodia. Accessed September 2, 2017. www.cambodiameteo.com/articles?menu=115&lang=en.

Missbach, Antje. *Troubled Transit: Asylum Seekers Stuck in Indonesia*. Singapore: EAS, Yusof Ishak Institute, 2015.

Mittelman, James H. *Hyperconflict : Globalization and Insecurity*. Stanford: Stanford University Press, 2010.

Mohanty, Samarendu. "The Global Rice Market Winks at El Niño and Thai Problems." *Rice Today* 13, no. 3 (July-September 2014). International Rice Research Institute (IRRI). http://irri.org/rice-today/the-global-rice-market-winks-at-el-nino-and-thai-problems.

Moses, Kara. "Lack of Forest Definition 'Major Obstacle' in Fight to Protect Rainforests." *The Ecologist,* October 20, 2010.

Murdiyarso, Daniel, Sonya Dewi, Deborah Lawrence, and Frances Seymour. "Indonesia's Forest Moratorium: A Stepping Stone to Better Forest Governance?" Bogor, Indonesia: Center for International Forestry Research, 2011.

"Myanmar: The Politics of Rakhine State." Asia Report No. 261. Brussels: International Crisis Group, October 22, 2014.

Nadaraj, Vanitha. "Levy for Foreign Workers in Malaysia: Who Will Blink First?" *ASEAN Today,* February 4, 2016.

National Academy of Sciences. *The Domestic and International Impacts of the 2009 H1N1 Influenza A Pandemic: Global Challenges, Global Solutions Workshop Summary.* Washington, DC: National Academies Press for Institute of Medicine, 2010.

Nesadurai, Helen E. S. "Malaysia's Conflict with the Philippines and Indonesia over Labour Migration: Economic Security, Interdependence and Conflict Trajectories." *Pacific Review* 26, no. 1 (2013): 89–113.

"News Release: SBY Vows to Protect Indonesia's Rainforests." *Jakarta Globe,* September 27, 2011.

"Nine Months In, How Has Jokowi Fared on the Environment?" *Conservation News,* July 23, 2015.

"1995 Mekong River Commission Agreement." Mekong River Commission, April 5, 1995.

Nugraha, Panca. "Gov't Urged to Reconsider Ban on Migrant Domestic Workers." *Jakarta Post,* February 15, 2015.

Olson, Mancur. *The Logic of Collective Action*. Cambridge: Harvard University Press, 1965.

Organisation for Economic Co-operation and Development (OECD). "Share of ODA for Agriculture (percent)." Paris: OECD Creditor Reporting System, 2009.

Osborne, Milton E. *River at Risk: The Mekong and the Water Politics of China and Southeast Asia*. Sydney: Longueville Media for Lowy Institute for International Policy, 2004.

Osman, Nurfika. "Jakarta, Riyadh to Sign MoU on Migrants." *Jakarta Post,* February 19, 2014.

Palmer, Wayne. *Indonesia's Overseas Labour Migration Programme.* Boston: Brill, 2016.

Parameswaran, Prashanth. "Malaysia Comes Under Fire from Indonesia Ahead of Jokowi Visit." *The Diplomat,* February 15, 2015.

Paris, Roland. "Human Security: Paradigm Shift or Hot Air?" *International Security* 26, no. 2 (2001): 87–102.

Pearce, Fred. "A Global Treaty on Rivers: Key to True Water Security." *Yale Environment 360,* November 19, 2012.

Pech, Sokhem, and Himesha Ranamukhaarachchi. "Water Governance Systems in Cambodia, Lao PDR and Vietnam." Vientiane, Laos: Mekong Program on Water Environment and Resilience, M-Power, 2013.

Piper, Nicola. "Migrant Labor in Southeast Asia." Singapore: Friedrich Ebert Stiftung Project on Migrant Labor in Southeast Asia, Asia Research Institute, National University in Singapore. http://www.fes.de/aktuell/focus_interkulturelles/focus_1/documents/8_000.pdf.

Purnomo, Agus. "Environment Caught in the Extremes." *Jakarta Globe,* April 19, 2011.

"Regional Overview of Food Insecurity in Asia and the Pacific." Rome: FAO, 2005.

"RI Migrant Workers Report Extortion to President Jokowi." *Jakarta Post,* November 30, 2014.

"Rice Cartel Plan Resurfaces." Radio Free Asia, March 21, 2012.

"Rice Shipments from Vietnam Reach 2.87 MMT to June 14th." Bloomberg News, June 17, 2012.

Richards, Daniel R., and Daniel A. Friess. "Rates and Drivers of Mangrove Deforestation in Southeast Asia, 2000–2012." *Proceedings of the National Academy of Sciences* 113, no. 2 (2016): 344–349.

Rieker, Pernille. "From Common Defense to Comprehensive Security: Towards the Europeanization of French Foreign and Security Policy?" *Security Dialogue* 37, no. 4 (2006): 509–528.

Riggs, Peter. "Indonesia's Colossal, Vital, and Unusual Role in the Global Climate Change Debate." Paper presented at the United States–Indonesia Society (USINDO) Open Forum, Washington, DC, May 4, 2010.

Rondonuwu, Olivia. "Indonesia Divided over Forest Moratorium." Reuters, January 5, 2011.

Rosenau, James N. "Governance, Order, and Change in World Politics." In *Governance Without Government: Order and Change in World Politics,* edited by James N. Rosenau and Ernst-Otto Czempiel, 1–29. Cambridge: Cambridge University Press, 1992.

Rukmana, Deden. "Recent Floods in Jakarta Are Strong Evidence that Urbanization Must Be Reduced." *Indonesia's Urban Studies Blog,* February 21, 2007. http://indonesiaurbanstudies.blogspot.com.

S. Rajaratnam School of International Studies (RSIS). "The 19th ASEAN Summit: Tackling Floods, Food And Stability—Analysis." *Eurasia Review,* November 17, 2011.

S. Rajaratnam School of International Studies (RSIS) Centre for Non-Traditional Security (NTS) Studies. *International Conference on Asian Food Security 2011—Feeding Asia in the 21st Century: Building Urban-Rural Alliances (10–12 August 2011).* Singapore: RSIS NTS, August 12, 2011.

Safman, Rachel M. "The Political Economy of Avian Influenza in Thailand." Working Paper No. 18. Brighton: STEPS Centre, 2009.

Sagita, Dessy. "Indonesia's Health Ministry Gives WHO Access to Bird Flu Samples Again." *Jakarta Globe,* April 27, 2015.

Saloh, Yani. "Moratorium Isn't the End of Business." *Jakarta Globe,* July 14, 2011.

Santi, Natalia. "Jokowi Pledges to Protect Indonesian Workers in Malaysia." *Tempo,* February 7, 2015.

SarDesai, D. R. *Vietnam: Past and Present.* Boulder, CO: Westview Press, 1998.

Sari, Agus P, Martha Maulidya, Ria N. Butarbutar, Rizka E. Sari, and Wisnu Rusmantoro. "World Bank: Indonesia and Climate Change Executive Summary."

Washington, DC: Department for International Development, International Bank for Reconstruction and Development (IBRD), 2013.

Satriastanti, Fidelis E. "Jokowi Bans New Oil Palm and Mining Concessions." Eco-Business, April 18, 2016. www.eco-business.com/news/jokowi-bans-new-palm-and-mining-concessions/.

Scheffran, Jürgen, and Antonella Battaglini. "Climate and Conflicts: The Security Risks of Global Warming." *Regional Environmental Change* 11, no. 1 (2011): S27–S39.

"Second Protocol to Amend the Agreement on the ASEAN Food Security Reserve." Joint Press Statement of the Tenth AMAF Plus Three, Phnom Penh, October 24, 2010.

Seymour, Frances, Nancy Birdsell, and William Savedoff. "The Indonesia-Norway REDD+ Agreement: A Glass Half Full." Policy Paper No. 56. Washington, D.C: Center for Global Development, February 2015.

Siegel, Robert. "How Fear Drove World Rice Markets Insane." NPR, November 2, 2011.

Sim, Shuan. "Thailand Rice Subsidy Scheme: What It Is and How It Toppled Thai Leader Yingluck Shinawatra." *International Business Times,* January 23, 2015.

Simamora, Adianto. "Govt Torn Between Courting Business, Protecting Forests." *Jakarta Post,* May 24, 2011.

"Simply REDD: CIFOR's Guide to Forests, Climate Change and REDD." Bogor, Indonesia: CIFOR, 2009.

"Singapore, Indonesia Jostle over Anti-Haze Measures." *Conservation News,* May 16, 2016.

"Sixty-fourth World Health Assembly Closes After Passing Multiple Resolutions." Geneva: WHO, May 24, 2011.

Slayton, Tom, and C. Peter Timmer. *CGD Notes: Japan and China Can Solve the Rice Crisis—But US Leadership Is Needed.* Washington, DC: Center for Global Development, 2008.

Smith, Alison. "Gone Fishing." *Southeast Asia Globe,* January 24, 2014.

———. "How Mekong Dams Could Affect Cambodia's Food Security." *Southeast Asia Globe,* January 24, 2014.

Snyder, Jack L. *From Voting to Violence: Democratization and Nationalist Conflict.* New York: Norton, 2000.

So, Grace. "Rice Pork and More to Come from Jilin Food Zone." *Business Times,* March 3, 2016.

Soriano, Zelda. "Climate Change Demands that ASEAN's Future Is Cooperative." *The Nation,* May 1, 2013.

Soroos, Marvin S. *The Endangered Atmosphere: Preserving a Global Commons.* Columbia: University of South Carolina Press, 1997.

"Southeast Asia Subregional Report, Asia-Pacific Forestry Sector Outlook Study." Bangkok: FAO, 2011.

Steni, Bernadinus. "Climate Change and ASEAN." *Jakarta Post,* May 10, 2011.

"Strategies Against Flu Emergence (SAFE) Project (2011–2013)—Ongoing." Washington, DC: US Agency for International Development (USAID) and US Department of State.

"Strengthening Health Security by Implementing the International Health Regulations." Geneva: WHO, 2005.

Suharmoko, Aditya. "SBY Vows Direct Oversight of Norway-Financed Projects." *Jakarta Post,* May 30, 2010.

Sunchindah, Apichai. "The Lancang-Mekong River Basin: Reflections on Cooperation Mechanisms Pertaining to a Shared Watercourse." Working Paper Policy Brief PO13-01. Singapore: RSIS NTS, February 2013.

Tabuchi, Hiroko, and Henry Fountain. "Bucking Trump, These Cities, States, and Companies Commit to Paris Accord." *New York Times,* June 1, 2017.

Teng, Paul, Mercedita Sombilla, J. Ewing, and Margarita Escaler. "Feeding Asia in the 21st Century: Building Urban-Rural Alliances: Summary of the Main Findings of the International Conference on Asian Food Security Held in Singapore on 10–12 August 2011." Singapore: RSIS, March 2012.

Teoh, Shannon. "Malaysia, Indonesia to Appoint Special Envoys to Handle Territorial Disputes." *The Straits Times,* February 6, 2015.

Tiensen, Thanawat., Prasit Chitaweesub, Thaweesak Songserm, Arunee Chaisingh, Wirongrong Hoonsuwan, Chantanee Buranathai, et al. "Highly Pathogenic Avian Influenza H5N1." *Emerging Infectious Diseases* 11, no. 11 (2004): 1664–1672.

Tiezzi, Shannon. "Facing Mekong Drought, China to Release Water from Yunnan Dam." *The Diplomat,* March 16, 2016.

Timmer, C. Peter. "Food Security in Indonesia: Current Challenges and the Long Run Outlook." Working Paper No. 48. Stanford: Center for Global Development, November 2004.

Timmer, C. Peter, and Thom Jayne. "Managing Food Price Volatility: Approaches at the Global, National, and Household Levels." In *Center on Food Security and the Environment: Stanford Symposium Series on Global Food Policy and Food Security in the 21st Century.* Stanford: Stanford University, May 26, 2011.

Tolarova, Katerina. "Global Governance: Fragmentation, Stagnation, Transformation? A Talk with Miles Kahler." University College London, December 13, 2016.

Torti, Jacqueline. "Floods in Southeast Asia: A Health Priority." *Journal of Global Health* 2, no 2 (2012): 1–6.

"Transboundary: Mobilising Grassroots Engagement and Facilitating High-Level Dialogue for Transboundary Water Management in the Mekong River Basin (#449)." Stockholm, Sweden: Global Water Partnership, June 16, 2014.

"The Trouble with Indonesia's Foreign Policy Priorities Under Jokowi." *The Diplomat,* January 9, 2015.

Tyuen, Le Danh. "Food and Health Security in Southeast Asia." *Asia Pacific Journal of Clinical Nutrition* 18, no. 4 (2009): 493–497.

"Under-Fire Jokowi Prepares Biggest Shift Yet in Indonesian Haze." *Mongabay Haze Beat,* October 26, 2015. http://news.mongabay.com/2015.

United Nations. "The Paris Agreement: Status of Ratification." United Nations Framework Convention on Climate Change, October 5, 2016. http://unfccc.int/2860.php

United Nations High Commissioner for Refugees. "UNHCR and Malaysia Close Camp for Vietnamese Boat-People." June 25, 2001. www.unhcr.org/en-us/news/press/2001/6/3ae6b81838/unhcr-malaysia-close-camp-vietnamese-boat-people.html.

———. "UNHCR Global Trends 2015." 2015. www.unhcr.org/en-us/statistics/unhcrstats/576408cd7/unhcr-global-trends-2015.html.

US Department of State. "Thailand: 2014 Trafficking in Persons Report." https://www.state.gov/j/tip/rls/tiprpt/countries/2014/226832.htm.

US Environmental Protection Agency. "Climate Change Basics." Updated January 19, 2017. www3.epa.gov/climatechange/basics/www3.epa.gov/climatechange/basics.

US Institute of Medicine Forum on Microbial Threats. *Workshop Overview.* Washington, DC: National Academies Press for Institute of Medicine, 2010.

"U.S. Relations with Indonesia." Washington, DC: US Department of State, January 17, 2017.

"USDA Foreign Agricultural Service Grain Report: Indonesia Rice and Corn Update." Washington, DC: USDA, April 23, 2010.

"USDA Post Estimates for Malaysia's Rice Production, Imports," *Oryza,* March 3, 2016.

Vidal, John. "Greater Mekong Countries 'Lost One-third of Forest Cover in 40 Years.'" *The Guardian,* May 2, 2013.

"Vietnam to Cooperate with Thailand on ASEAN Rice Cartel." *Angimex,* June 26, 2012. http://www.angimex-kitoku.com.vn/en-US/home.

Vongkiatkajorn, Kanyakrit. "The Year Southeast Asia Faced Its Own Refugee Crisis." *VICE News,* December 27, 2015.

Water Rights and Water Allocation: Issues and Challenges for Asia. Los Baños, Philippines: Asian Development Bank, 2009.

Weng, Lawi. "Ma Ba Tha Leader Tells Monks to Support Incumbent Govt." *The Irrawaddy,* June 30, 2015.

Westing, Arthur. "The Environmental Component of Comprehensive Security." *Bulletin of Peace Proposals* 20, no. 2 (1989): 129–134.

Wishnick, Elizabeth. "Dilemmas of Securitization and Health Risk Management in the People's Republic of China: The Cases of SARS and Avian Influenza." *Health Policy and Planning* 25, no. 6 (2010): 454–466.

Wolf, Aaron T., and Joshua T. Newton. "Case Study of Transboundary Dispute Resolution: The Mekong Committee." Eugene: Oregon State University Program in Water Conflict Management and Transformation, 2000.

Wong, Edward. "The Key Players in Climate Change." *New York Times,* April 21, 2016.

Workman, Daniel. "Rice Import Dependent Countries Nigeria, Saudi Arabia and Philippines Among Biggest Rice Buyers." Los Baños, Philippines: International Rice Research Institute, April 16, 2008.

"World Health Organization, Avian Influenza." New Delhi: World Health Organization Regional Office for Southeast Asia, (SEARO), 2013.

"Worries Grow for Rohingya, Bangladeshi Migrants as Malaysia, Indonesia Refuse to Accept Them." Associated Press, May 13, 2015.

"Xayaburi Dam." Berkeley, CA: International Rivers Organization, April 2014.

"Xayaburi Dam Timeline of Events." Berkeley, CA: International Rivers Organization, April 2014.

Xaypaseuth, Phomsoupha. "Hydropower in Lao PDR—Department of Energy Business." Paper presented at the conference "Power in Progress," Vientiane, Laos, March 2016.

Zawahri, Neda A., and Sara McLaughlin Mitchell. "Fragmented Governance of International Rivers: Negotiating Bilateral Versus Multilateral Treaties." *International Studies Quarterly* 55, no. 3 (2011): 835–858.

Zubiada, Nazeer. "Indonesia Loses Forests as Politicians Bicker." *Jakarta Globe,* April 19, 2011.

Zulfakar, Mergawati. "Jokowi Visit a Chance to Manage Ties." *The Star Online,* February 5, 2015. www.thestar.com/my/opinion/columnists/mergawati/2015/02/05/jokowi-visit-a-change-to-manage-ties/.

Zweynert, Astrid. "Indonesia's Maid Ban Increases Risk of Human Trafficking: Campaigner." Reuters, June 17, 2015.

Index

About the Book

With the countries of Southeast Asia increasingly challenged by a plethora of nontraditional security issues—climate change, food and water security, infectious diseases, and migration key among them—a number of important questions have emerged: What national and regional efforts are being made to address these issues? Why have some approaches proven more successful than others? How do competing private and public interests affect the ability of states to protect their citizens?

Addressing these questions, Amy Freedman and Ann Marie Murphy explore the factors that continue to impede cooperation in combating transnational security threats across the region.

Amy L. Freedman is professor of political science and international studies at Long Island University Post. Her previous books include *Threatening the State: The Internationalization of Internal Conflicts*. **Ann Marie Murphy** is associate professor in the School of Diplomacy and International Relations at Seton Hall University. She is coeditor of *Legacies of Engagement in Southeast Asia*.